Cane toads

A tale of sugar, politics and flawed science

Nigel Turvey

SYDNEY UNIVERSITY PRESS

First published in 2013 by Sydney University Press

© Nigel Turvey 2013
© Sydney University Press 2013

Reproduction and Communication for other purposes

Except as permitted under the Act, no part of this edition may be reproduced, stored in a retrieval system, or communicated in any form or by any means without prior written permission. All requests for reproduction or communication should be made to Sydney University Press at the address below:

Sydney University Press
Fisher Library F03
University of Sydney NSW 2006
AUSTRALIA
Email: sup.info@sydney.edu.au

National Library of Australia Cataloguing-in-Publication Data

Title: Cane toads: a tale of sugar, politics and flawed science / Nigel Turvey.
ISBN: 9781743323595 (pbk)
          9781743323717 (ebook: epub)
          9781743323724 (ebook: kindle)
Notes: Includes bibliographical references and index.
Subjects: Bufo marinus--Australia.
          Bufo marinus--Control--Australia.
          Toads--Control--Australia.
          Biological invasions.
          Nonindigenous pests--Control--Australia.
          Sugarcane--Diseases and pests--Control--Australia.
Dewey Number:
          597.80994

Cover image by Mark Lewis, Radio Pictures, Mullumbimby
Cover design by Miguel Yamin

Contents

List of figures   v
Preface: of toads and men   ix

1 The ancestors – fossil 41159   1
2 The apothecary's toad   11
3 The sweet grass   25
4 Queensland sugar – Hope and Whish   37
5 Ladybird fantasy   53
6 The cane beetles   73
7 Hawai'i leads biological control   87
8 Birth of a myth   101
9 Toad fantasy   115
10 Toads for Queensland   125
11 War on Canberra   143
12 Living with Bufo   157
13 Cane toad wars   173
14 Taking the Top End   185
15 Bad, flawed and reckless   203

Works cited   219
Index   241

List of figures

Figure 0.1 Redistribution of Bufo marinus from South America. 10
Figure 0.2 Redistribution of Bufo marinus from Hawai'i. 11
Figure 2.1 Albertus Seba in his kunstkammer about 1734. 13
Figure 2.2 Toad in Seba's *Thesaurus* referred to by Carolus Linnaeus as *Rana marina* (Tomus 1 Tabula 76). 21
Figure 2.3 *Industria*. 23
Figure 3.1 Toad market in Paris 1879. 30
Figure 4.1 Captain The Hon. Louis Hope. 40
Figure 4.2 Walter Hill, Superintendent, Queensland Botanical Gardens. 41
Figure 4.3 John Buhôt. 43
Figure 4.4 Captain Claudius Buchanan Whish. 45
Figure 5.1 Charles Valentine Riley, Chief Entomologist, US Department of Agriculture. 54
Figure 5.2 Albert Koebele, HSPA entomologist. 56
Figure 5.3 Vedalia beetle eating cottony cushion scale. 59

Figure 5.4 Robert Perkins, HSPA entomologist.	65
Figure 5.5 Frederick Muir, HSPA entomologist.	68
Figure 6.1 Alan Dodd.	76
Figure 6.2 Larva, pupa and adult greyback cane beetle.	81
Figure 6.3 Prickly-pear forest 1930.	82
Figure 7.1 Cyril Pemberton, daughter Virginia and Indian motorbike 1916.	88
Figure 8.1 *Bufo marinus* pair in amplexus.	106
Figure 8.2 Dissection of the stomach of *Bufo marinus*. Bulburin, Queensland.	108
Figure 8.3 Stomach contents of *Bufo marinus*. Bulburin, Queensland.	109
Figure 9.1 Cyril Pemberton, HSPA entomologist with *Bufo marinus*.	119
Figure 10.1 Arthur Bell, BSES.	128
Figure 10.2 Bill Kerr, BSES.	131
Figure 10.3 *SS Mariposa* departing Sydney Cove March 1932.	136
Figure 10.4 Reg Mungomery, BSES.	139
Figure 11.1 Walter Froggatt at left, captioned 'Scientific staff of the Sugar Planters Association Honolulua. Chambers, Kotinski, Swezy, Davis', c. 1908.	144
Figure 11.2 Walter Froggatt, Linnean Society of New South Wales, 1923.	145
Figure 11.3 The Hon. Frank William Bulcock, 1942.	151
Figure 11.4 Queensland Department of Agriculture and Stock, 95 William Street, Brisbane, backing onto Queen's wharf on the Brisbane River, June 1936.	152
Figure 12.1 Monica Krauss with her pet cane toad Dairy Queen, Queensland, 1984.	166

List of figures

Figure 14.1 Ben Phillips. 190

Figure 14.2 Professor Rick Shine. 199

Preface: of toads and men

January 1532, estuarine mudflats of Baia de Santos on the coast of Brazil. Portuguese sailors rowed ashore on a flooding tide, breached mangrove barricades and landed their commander Martim Afonso de Sousa, Governor of the Land of Brazil.[1] In the Bay of Saints he sought blessings from on high and contemplated the magnitude of his tasks: chase away the French, harvest *pau brasil* (brazilwood [*Caesalpinia echinata*]), plant sugar cane and found a nation. Saints were beseeched, forests cleared, soils tilled, billets of cane trimmed, laid in rows and buried. The giant toad *aguaquaquan*, known also as *Bufo marinus, Rhinella marina* and now colloquially as the cane toad, would look on as sugar cane plantings spread northwards into its homeland.

January 2013, wet season, northern Australia. A savannah of sparse trees and resinous grasses scavenged sustenance from stony hills in the East Kimberley Ranges, Western Australia. A stately goanna flicked its forked tongue under a fire-blackened log and sensed food, sensed the vanguard of cane toads heading ever westwards across Australia's tropical north, nocturnal invaders waiting out the heat of the day in the shade of the log. The goanna ate the toad in two swallows of its long neck, staggered a while, regurgitated a mucous lump, collapsed

---

1   Augeron & Vidal 2007, p. 23.

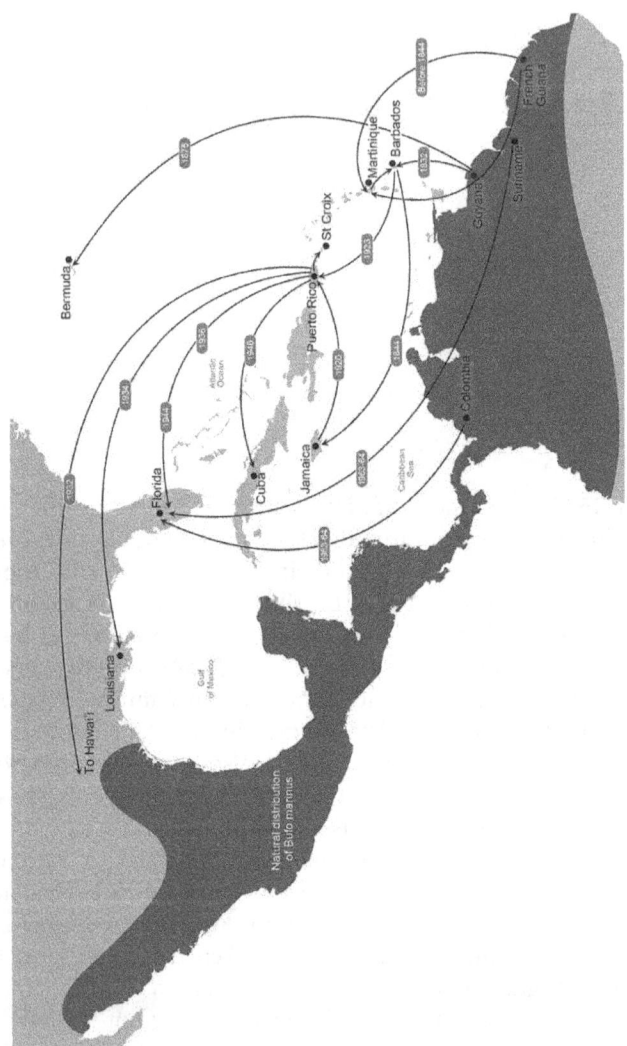

Figure 0.1 Redistribution of Bufo marinus from South America. (Redrawn and updated after Easteal 1981 and Zug & Zug 1979.)

Preface: of toads and men

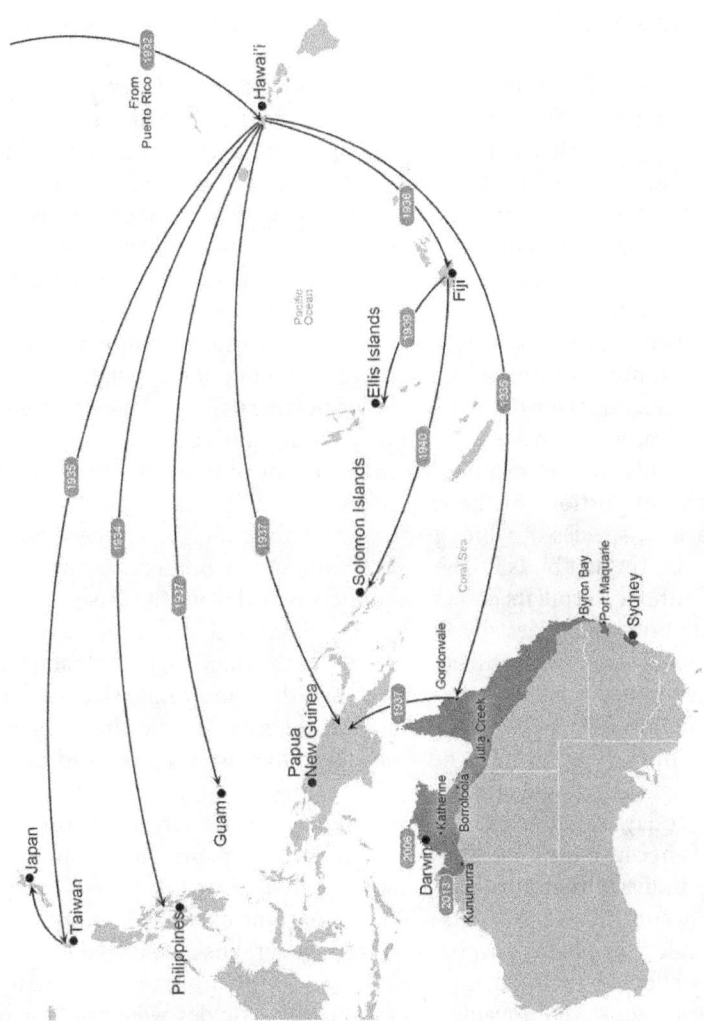

Figure 0.2 Redistribution of Bufo marinus from Hawai'i. (Redrawn and updated after Easteal 1981 and multiple sources.)

and died. Australia's native fauna are unused to the toxins of foreign toads.

These two seemingly unrelated events on different continents bracket almost 500 years of history for sugar cane and the cane toad. Over almost half a millennium, cane toads, assisted by generous mankind, sugar cane farmers, scientists and the fortunes of war, colonised some 138 countries and islands,[2] ending up as far from home as Florida, Bermuda, Hawaii, Fiji, Guam, the Philippines and Japan. And if the introduction of sugar cane to the toad's homeland was the Yin, the Yang was the introduction of cane toads to the island of Papua New Guinea, the ancestral home of sugar cane. Cane toads now occupy most of the tropical north of Australia – the largest contiguous population of cane toads outside their natural habitat of northern South America, Central America, southern Mexico and neighbouring islands.

Toads were introduced by man to control pests in sugar cane because 'all portions of the cane plant are subject to destructive insect attack ... species of white grubs devour the roots ... wire worms may eat into the sett buds ... the stem is attacked by borers ... army worms and other caterpillars and grasshoppers may devour the leaves ... while moth borers may destroy the growing point of the cane'.[3] In sugar cane plantations on the northern coast of South America, native cane toads were reputed to eat so many pests that 19th-century English and French plantation owners carried them from the mainland to their sugar estates in the Caribbean. And from there they were introduced around the sugar growing world.

Today, it is hard to understand why toads were used to control pests because now we simply pick up a can of spray to kill an insect. The industry that produces today's synthetic insecticides is a 'child of the Second World War'.[4] Its siblings, the chlorinated hydrocarbons and organic phosphates now banned from general use, delivered spectacularly effective pest control – at a cost to the environment. Before these new poisons were invented, agricultural pesticides were crude, nasty, and ineffective products like tar, copper, sulphur, lead and arsenic compounds. Without effective pest control, farmers and gardeners stim-

---

2   Lever 2001.
3   Kerr & Bell 1939, p. 172.
4   Carson 1968(1962), p. 31.

ulated a worldwide currency in agents of 'biological control'. These agents included such 'beneficial' organisms as carnivorous ladybirds, parasitic wasps and flies, insectivorous birds, cactus-hungry moths, voracious mongoose and omnivorous toads. Sugar cane scientists were among the busiest traders in this currency but, although these agents were biological, their promoters had no control over their behaviour.

The emergence of the cane toad as an agent of biological control has its foundations in the unregulated distribution and release of 'beneficial' organisms. That is why this story details the parade of beetles, wasps and flies that were released, untested, into sugar cane fields; it helps us to understand the actions of scientists responsible for the later release of the cane toad into unsuspecting environments.

In 1935 sugar scientists released the cane toad into sugar cane fields in Queensland on Australia's tropical east coast. The release was supported by cane growers, leading scientists, and politicians including the prime minister of Australia. At the time, none but a lone voice thought it possible that an exotic amphibian could traverse this continent's arid outback – often too tough for men – let alone breed and colonise new environments. It took 74 years for toads to traverse Australia's tropical north to arrive at the Western Australian border. Generation after generation of toads hopping in an evolutionary relay, grandparent to parent to offspring, over and again, more than 2,400 km across hot and mostly arid landscapes.

Cane toads failed spectacularly in their allotted task in the cane fields. Their task was to eat the adults of soil-dwelling beetle larvae that ate the roots of sugar cane, and thereby control populations of the pest. They failed in Caribbean islands, in Hawai'i and in Australia. Cane toads will eat almost anything that moves and is small enough to swallow, but their reputation in targeted pest control was gained through erroneous deduction, desperation and wildly optimistic extrapolation. To control populations of soil-dwelling larvae by eating airborne beetles is a tall order for any predator no matter how hungry. Cane toads could never control populations of the beetles they were tasked with.

The great leaps forward in global distribution of toads were done with the help of men and women bereft of ideas about how else to control the pests that were eating their sugar cane or garden crops. They were well respected and well intentioned figures in their communities, not villains out to colonise the country with a toxic alien amphibian. But their stories are a litany of desperation – a desperation that caused myopia. And myopia caused errors that were compounded infinitely by the extraordinary breeding and colonising capabilities of the cane toad. The resulting invasion of many territories, including northern Australia, is a sad reflection on those who championed the toad, on the times, the state of science and the paucity of quarantine procedures that allowed its importation and release.

This story is about good intentions that turned bad. It is a tale of good-ideas-at-the-time, a tale of unintended consequences and, most of all, a tale of simple acts leading to catastrophic outcomes. This story about human nature has modern parallels. It is about scientists so committed to helping solve a problem that they are blinkered from viewing adverse impacts; committed to serving their country, their leaders and their benefactors – the industry that employed them. There is no blame to apportion – just lessons to learn from the toad's tale. And, as the tale shows, we come perilously close to repeating the mistakes of the past.

1
The ancestors – fossil 41159

Before thought, before reason, before mankind, lived a toad singular among peers. It squatted motionless, by a swamp. Hind legs folded, warty brown torso propped on stumpy forelegs, hands turned inwards, broad beak-like head thrust forward. Eyelids blinked each yellow iris.

Afternoon thunder rolled into evening, drip tips on rainforest leaves shed the storm's remnants. Mist lay over the swamp. Without a ripple, a caiman's triangular snout parted the waters, evaluated prey, and waited. In wet canopies, monkeys disputed hierarchies. Squabbles shed showers onto a four-metre-long armadillo, its armour glistened as it foraged the forest floor.

Large water drops alerted the toad. A beetle, intent on purpose, traversed wet grass and leaves, climbed up, tumbled down, climbed again. The toad lunged, snapped up the beetle, ran rounded tongue around thin lips. The river rose a little, swollen by runoff from the hills, quietly broached banks, gently flushed swamps on the flood plain. This toad, father of fathers, surrounded by rising waters, swam a while, struggled for firm ground, then drowned. The river fell, left fine silt on the flood plain and entombed the toad – for the next 15 million years.

April 1949, La Venta in the Tatacoa desert in the upper reaches of the Magdalena River of northern Colombia. Twenty-nine year old Bob Fields, one-time supermarket manager, US Army Air Force sergeant, now palaeontologist, found the toad. It was fossilised, embedded in fine sediments in the Villavieja Formation at La Venta. He numbered the fossil RWF191 and added acquisition 41159 to the register of the University of California's Museum of Palaeontology.[1]

This singular toad became a numbered toad, a reference for its kin.

The deserts of the upper Magdalena Valley lie in the rain shadow, east of Colombia's Central Cordillera, just north of the equator. Occasional rainfall erodes soft sediments sparsely protected by vegetation – the badlands. Weirdly dissected, constantly changing, deep ravines are flanked by precarious pillars, capped by solitary cacti. Snakes, scorpions and spiders share the hot dry landscape and shelter from freezing nights in crevices. Fields learnt his first foreign language around a desert campfire. The language was a Spanish patois strengthened with expletives; his teachers were itinerant Colombian labourers.[2]

It was a good expedition. Sediments more than a kilometre deep trapped fossils. They were from the Miocene, between 5 million and 23 million years ago. Fields's team dug into sediments, scrape by scrape. He called it the Monkey Field because of fossils of new world monkeys, one later named after Fields – *Neosaimiri fieldsi*.[3] There were fossils of giant creatures, now extinct: a glyptodont – an armadillo the size of a small car – a giant iguana, a ground sloth, giant snorkelled grazing animals, giant rodents. And the toad, fossil 41159, unremarkable among extinct giants.

Summer 1963, the basement of the Harvard Museum of Comparative Zoology. Eighteen year old Richard Wassersug stared at fossil 41159. This high school student at Thayer Academy, Boston, received a National Science Foundation grant to move among great men, to stimulate a passion for science – a result of the space race with the Soviet Union. Guided by Professor Richard Estes, young Wassersug compared fossil 41159 with skeletons of modern amphibians. On the left forearm a raised ridge, the *crista medialis*, defined it as male. A strong muscle attaches to the ridge and flexes the wrist to grasp a female for mating, called amplexus. It was almost identical to the toad *Bufo marinus*.[4] Fossil 41159 looked remarkably modern – or the modern skeleton remarkably ancient.

1　Holroyd 2009.
2　Weidman 1997.
3　Holroyd 2009.
4　Estes & Wassersug 1963.

1 The ancestors – fossil 41159

Bufo marinus, a bufonid,[5] a true toad, still hops around the flood plains of the Magdalena. Bufonids branched off from South American frogs around 88 million years ago,[6] hopped into North America, crossed land-bridges into Asia, hopped down into Europe, across to Africa, and left groups behind to adapt and become new toad species.

Around 65 million years ago, the toads' travels only half done, an asteroid hit earth. Dust encircled the globe, blocked out the sun, caused mass mortality – the K-T extinction. Dinosaurs died out: seven tonne *Tyrannosaurus rex*, flying dinosaur *Quetzalcoatlus* with its 10 metre wing span, and 17 metre long marine reptile *Hainosaurus*.

Toads hopped on.

Around 43 million years ago, new toad species hopped from Asia back across the top of the world into North America and down into South America. Some evolved as *Rhinella* or beaked toads. Out of these evolved *Rhinella marina*, the new name for *Bufo marinus* since 2008.[7] But since most of this story and the characters who promoted the toad pre-date the name change, in the narrative of the toad's tale we will stay with the familiar name *Bufo marinus*.

Another mass extinction occurred 10 thousand years ago, just yesterday for geologists. Its cause more sinister – the spread of man and dramatic changes in climate. Many species disappeared from the Americas: sabre toothed cat *Smilodon*, giant ground sloth *Megatherium*, two metre long beaver *Castoroides*, mammoths and mastodons. Flightless birds, three tonne wombat *Diprotodon* and marsupial lion *Thylacoleo carnifex* vanished from Australia. Giant deer *Megaloceros* with three metre wide antlers, woolly mammoth, woolly rhinoceros, cave lion and cave bear vanished from Europe and Asia.

Toads hopped on.

Extinctions helped. On their journey toads clustered around carrion, waited for insects, then snapped them up, as toads do today.

---

5 Frogs and toads are called anurans, meaning without tails, but frog and toad are not terms used in taxonomic classification. True toads are called bufonids. Ugly, warty frogs are sometimes called toads but they are not bufonids. While all bufonids are toads, not all anurans called toads are bufonids.
6 Pramuk et al. 2008.
7 Pramuk et al. 2008.

Fifteen million years ago, bufonids covered the Americas, Europe, Africa, and Asia. But the island continent of Australia, islands in the Caribbean and emerging volcanic islands like Hawai'i, remote in their oceans, remained free of toads. *Bufo marinus* established its natural range: northern South America, Central America, southern Mexico and close-by islands – Trinidad, Tobago, Little Tobago, Archipelago de Las Perlas and Isla de Cozumel.

Bufo marinus is a survivor. It retains ancient traits from when amphibians first colonised land. As climates on land became drier, amphibians used seasonal pools of water for breeding. In puddles their eggs grew rapidly into tadpoles. Tadpoles changed quickly into land dwelling adults and moved away to find shelter as pools dried up.[8] It is a colonising strategy still highly effective today. The tadpole, a small-mouthed water-dwelling larva, has gills, a tail, and no reproductive organs. Within 10 to 20 days, the tadpole changes to a land-dwelling, air-breathing, carnivorous, tailless, long-legged, reproductive amphibian.[9] It is a remarkable transformation. A relic of the adaptive transition from sea to land, unchanged, frozen in time.

1970, Costa Rica, the University of California field school. Richard Wassersug, a little older, wondered about the behaviour of tadpoles. Most tadpoles hide when threatened but *Bufo marinus* is brazenly different. Its tadpoles form conspicuous shoals of many hundreds of individuals. Wassersug formed a hypothesis that highly visible tadpoles, like *Bufo marinus*, have no need to hide because they are unpleasant to eat. Predators will avoid them. He gathered eight species of tadpoles and 11 colleagues with a mean age of 27.5 years. No cigarettes to corrupt tastebuds, but beer provided because of the methodology: 'A tadpole was rinsed in fresh water. The taster placed the tadpole into his or her mouth and held it for 20–30 seconds without biting into it. Then the taster bit into the tail, breaking the skin and chewed lightly for 10–20 seconds. For the last 10–20 seconds the taster bit firmly and fully into the body of the tadpole.'[10] That's when beer was needed.

Someone kept a record – *Bufo marinus* left a bitter taste. Despite being highly visible in pools and vulnerable to attack, *Bufo marinus* was

---

8   Slade & Wassersug 1975.
9   Wassersug 1975.
10 Wassersug 1971; Wassersug 2008.

the least palatable and had no need to hide. And the most palatable tadpole was *Smilisca sordida*, the drab tree frog whose adults have big eyes, smooth green-banded skin and long back legs – a Kermit frog.

Despite the methodology and the beer, this is science. Questions were induced from field observations (Why did some tadpoles behave oddly?). A hypothesis was deduced (that they were unpleasant to eat). A replicated experiment (eight tadpole species replicated across 11 testers) provided a conclusion (conspicuous tadpoles were unpalatable). In the year 2000, the now bearded bearded Professor Wassersug received an Ig Nobel prize for 'scientifically minded achievements that cannot or should not be reproduced'.[11] He noted on reflection, 'Obviously, it's unethical to bribe graduate students with beer to get them to be subjects in a research project – they already drink too much beer'.[12]

April 1974, Panama, Barro Colorado Island, site of the Smithsonian Tropical Research Institute. George Zug, Associate Curator of Amphibians and Reptiles, had a long journey from Washington. Planes, a train alongside the Panama Canal and a small boat linked him to his place of research. Zug journeyed between Washington and the island eight times between 1974 and 1976, twice with his research assistant, his wife Patricia. Zug and Zug,[13] their jointly authored monograph published by the Smithsonian Institute is a rare study of *Bufo marinus* in its native environment. On Barro Colorado the toad's preferred habitat is open ground with cover close by. It does not like the forest and continuous forest is a barrier to movement. Adults are heavy-bodied. The head broad with bony ridges, truncated snout, and oval parotoid glands above the shoulders near the distinct ear drum. Fingers are free and toes are distinctly webbed. Adult males are coloured cinnamon-brown with very warty skin on backs and legs, each wart capped by one or many horny spines. The Zugs described them as 'mobile cow patties'.

These cow patties are toxic. The parotoid, or parotid or parotic glands – near the ear – contain a cocktail of lethal poisons, one of which is called bufotenine. This means a toad poison that causes paralysis. The glands are a passive defence. The urban myth is wrong – cane toads cannot squirt poison autonomously. When a predator mouths the toad

---

11 *Harvard University Gazette*, 28 September 2000.
12 Anon 2000.
13 Zug & Zug 1979.

and puts pressure on the glands they weep the milky poison. If ingested, it disrupts the nervous system, causes hallucinations, stops the heart and ends in sudden death.[14] And in addition to poison glands, 'anuran skin contains a vast pharmacopoeia of neuroactive and vasoactive peptides'[15] – substances that affect both nervous and circulatory systems. The toad is toxic in all stages of its life cycle – eggs, tadpoles, adults, males and females. In its native South America, predators of *Bufo marinus* are inured to its poisons. And in Florida and in the West Indies, it is hunted by raptors[16] familiar with toads. They avoid its toxic glands and skin, eating only the viscera. But in new environments colonised by *Bufo marinus* these poisons are the biggest problem. For predators not evolved with toads it is a meal easy to catch but fatal to eat. Australian animals evolved in isolation from toads. For quolls, black snakes, death adders, goannas and freshwater crocodiles, toad toxins cause a painful quick death.

In its natural habitat *Bufo marinus* grows in length at around 0.4 mm per day, reaches sexual maturity at around 90 mm in length from snout to vent, and grows to between 100 and 150 mm long.[17] Males increase in size over one year. Females grow for three years and lay stringed clusters of around 30,000 eggs twice a year.

In new environments without pathogens and predators they reach remarkable sizes. A well-fed female can weigh 1.5 kilograms – the size of a small roasting chicken – and to most they are ugly!

Bufo marinus is not a fussy eater and 'will eat almost every animate object it can catch'.[18] Although when offered it would not eat small chickens,[19] it will eat mice, small birds, 'poi [fermented taro], boiled rice, cooked cream of wheat and Carnation [milk] flakes … grasshoppers, armyworms, beetles, earthworms, scorpions, wasps, moths, bees, house lizards or geckos, ants, many sorts of caterpillars, snails in the shell, cockroaches and sow bugs or slaters … [and] after swallowing such a fiery creature as a carpenter bee, *Bufo* was observed to execute a

---

14   Lever 2001.
15   Wassersug 1997.
16   Schwartz & Henderson 1991, p. 20.
17   Zug & Zug 1979.
18   Zug & Zug 1979, p. 24.
19   Dexter 1932.

few abdominal motions suggestive of the Hawaiian hula dance'.[20] They eat items around the home, household waste, cat food, dog food, and rehydrate in drinking bowls. They camp around beehives waiting for heavily laden workers to come within range of their tongues, forage through vegetable gardens for slugs and snails and camp on cow pats waiting for dung beetles to emerge. On the rare occasion that *Bufo marinus* swallows something irritating, it will eject stomach contents and stomach as well. *Bufo marinus* is right handed and right footed[21] and uses its right hand to wipe away the irritant – gastric grooming – and push the stomach back into its body,[22] then squat, waiting for the next meal to appear.

Toads apply simple logic to life: if it's big, avoid it; if it's small, eat it; if it's in between, mate with it.

Bufo marinus is tough. In its native habitat it carries intestinal roundworms (*Nematoda*), thorny headed worms (*Acanthocephala*) and flatworms (*Trematoda*).[23] It also carries parasitic ticks on its body.[24] It ceases to function below 10°C and tolerates up to 42°C.[25] It doesn't drink and is vulnerable to dehydration. To re-hydrate it must sit on something moist and absorb water through a patch of skin on its belly. It loses water rapidly[26] and can lose up to 53% of its body weight before dying.[27] It can move water around its body using channels under its skin and use its bladder to help regulate water.[28] It can also tolerate salinity, increasing the salinity of its own blood to match that of the external environment. Tales of *Bufo marinus* inhabiting hot, exposed saline mudflats on the coast of Suriname – told by Dutch sailors in the 18th century – are probably the reason why it was then named *Rana marina maxima*,[29] the great marine toad.

---

20   Pemberton 1934, p. 190.
21   Malashichev & Wassersug 2004.
22   Naitoh & Wassersug 1996.
23   Schwartz & Henderson 1991, p. 20.
24   Schomburgk 1848, p. 679.
25   Young et al. 2005.
26   Young et al. 2005.
27   Lever 2001, p. 14, citing Zug & Zug 1979.
28   Lever 2001.
29   Müsch et al. 2001.

What of outer space? Richard Wassersug, of Ig Nobel fame, took his research high above earth, conducting research on amphibians in NASA Space Shuttles and in microgravity on parabolic flights. In microgravity, tadpoles floated weightlessly above the blue planet but had problems knowing which way was up. Most adult amphibians did not get motion sickness[30] but *Bufo marinus* became confused, twisted its hind legs, and became helplessly tangled in a pirouette.[31]

At least outer space may limit the march of *Bufo marinus* through the universe.

Many fear ugly warty toads. They were associated with witches.[32] Shakespeare called them 'spotted' and 'loathsome'. Carolus Linnaeus, father of systematics, did not have much regard for them either. He believed their 'horrible cold bodies, filthy colour ... fierce faces, ponderous features ... raucous calls, squalid habitats, and dreadful venom [was the reason why] ... the Creator had not made many of them'.[33]

But some love toads. Queenslanders are especially proud. In the annual State of Origin tribal rugby league challenge, Queensland's team, the maroons, affectionately called the cane toads, defeated New South Wales's team, the blues, or cockroaches, eight years in a row from 2006. Through Kenneth Grahame's books, and later through film and television, generations of children came to love Toad of Toad Hall. In *The Wind in the Willows*, Ratty's opinion of a cross-dressing *Bufo bufo* was that he was 'so simple, so good-natured, and so affectionate'.[34] And toads make loyal pets with simple demands – food of any kind, moisture and shelter.

In its home range and in the Caribbean it is called aguaquaquan, kwapp, macao, maco pempen, maco toro, sapo buey and sapo grande.[35] Elsewhere it is variously called crapaud – generally meaning toad, Aga-Kröte, giant American toad, great Mexican toad, South American toad,

---

30   Wassersug et al. 1993.
31   Wassersug 2001.
32   Toads, cats and rats were depicted accompanying witches in *phantasia* drawings by Dutch artists such as Jaques de Geyn II and Peter Breugel (Swan 2005).
33   Linnaeus 1758, p. 194.
34   Grahame 1995(1908), p. 27.
35   Multiple sources including Global Invasive Species Database.

Central American toad and Suriname toad.[36] In Puerto Rico it was known as *Bufo marinus* and when taken to Hawai'i it was called simply bufo. When taken to Australia it was called giant American toad, shortened to giant toad, then Queensland toad – there were no other toads in Australia. Finally, around 1949, it was called cane toad because by then it had migrated out of Queensland's cane fields and become an established pest. Cane toad is now an almost universal appellation. The cane toad: huge, brown, warty, repugnant, loaded with poisons, tries to eat anything it can swallow, tolerates extremes of cold, heat, salinity and dehydration, breeds in vast numbers and is hard to kill. No matter what it is called, the cane toad, giant toad, *Bufo marinus*, toxic amphibian, survivor, invader, occupier, is here to stay, warts and all.

Modern man came out of Africa less than two hundred thousand years ago, evolved rapidly and put an indelible mark on the planet. But the design of *Bufo marinus* has likely changed very little in the last 40 million years. Fossil 41159 is evidence, if any were needed, of the cane toad's capacity for longevity and adaptability. It is a survivor and, with the help of man, a master of colonisation.

---

36   Not to be confused with *Pipa pipa*.

2
The apothecary's toad

1667, Suriname on the north coast of South America was ceded by the British to Dutch control after The Netherlands lost a brief war. Peter Stuyvesant, Director General of the Dutch West India Company (WIC), surrendered mosquito-infested Manhattan Island in exchange for Suriname sugar plantations, complete with native toads. The victorious Duke of York renamed New Amsterdam eponymously. The exchange helped revitalise the WIC and stimulate the Dutch trade in sugar, pirated valuables and exotic collectibles.

1732, Amsterdam, a *kroeg* – a waterfront tavern. In a corner lay a mound of small linen sacks, stitched offcuts of sailcloth tied with jute cord. Some were rounded, filled to the neck, some held large misshapen items, one was empty except for a weighted object ringed by a damp circle of sawdust. A salt-tanned bosun, the crew chief, commanded the corner. He was the self appointed broker of oddities snaffled on the voyage. Remnants of his crew slumped on oak benches, the worse for wear, but home at last and yearning for *hutspot* and *klapstuk*, beef and vegetables.

There was a throng in the *kroeg*. Warm air mingled sour clothing, stale beer and tobacco. Paid-off sailors, expansively generous, rubbed shoulders with thirsty stevedores and penniless sailors begging a passage or a tot of rum. Merchants gathered in conspiratorial clusters, closing deals. A sisterhood of rounded matrons and hollow-cheeked consumptives moved among tables bringing bread, ale and female comfort for those with a few stuivers, or shillings, to their name.

The door to the waterfront opened. The throng hushed. The doorway framed a silhouette, a gentleman bewigged under a broad brimmed black hat, swathed in a dark cloak of Bruges velvet. He stood erect and surveyed the room with an easy familiarity. A lace kerchief in a gloved hand, an ivory handled cane in the other. Sailors, stevedores, merchants, wenches instinctively bowed in respect.

He was the apothecary.

The apothecary was followed in train by a tousle-haired boy lugging a mahogany box, harnessed and buckled about him like a regimental side drum. The bosun shooed the throng, cleared a place at the table, warmly greeted his regular customer, congratulated him on his health and gave mock apologies for his own long absence at sea. The press of people allowed the apothecary the entire bench to himself.

Albertus Seba, the apothecary, had been trading on the waterfront for decades. He sought out sailors newly arrived from the East or West Indies for his business, *Die Deutsche Apotheke*, the German Apothecary's Shop.[1] He was 67 years old, active and healthy. That morning he had just one target, the bosun and crew of the newly arrived West Indiaman.

Seba's stock and trade depended on the weather. A brisk westerly had brought the West Indiaman into the Zuiderzee. But there it sat – *voor Pampus liggen*[2] – opposed by both wind and tide and unable to enter the brackish bay of the IJ leading to the port of Amsterdam. Hanging limply from its stern was the distinctive red, white and blue horizontal bands of a pennant, the Geoctroyeerde Westindische Compagnie, the WIC or West India Company. It was newly arrived from South America, loaded with a cargo of muscovado – raw sugar. So close to home yet outfoxed by nature, the captain had to negotiate the services of the *scheepskameel* – extra flotation – to cross the shallows and enter port.[3]

---

1   Albertus Seba was born in Friedeburg in Lower Saxony.
2   This phrase means to lie, frustratingly, becalmed in front of Pampus Island at the entrance to the port of Amsterdam. Colloquially, in Amsterdam, this phrase is also applied to someone who is drunk, or without means of direction.

2 The apothecary's toad

Figure 2.1 Albertus Seba in his kunstkammer about 1734. (Den Haag, Koninklijke Bibliotheek. [Locupletissimi rerum naturalium thesauri accurata descriptio Naaukeurige beschryving van het schatryke kabinet der voornaamste seldzaamheden der natuur/Albertus Seba.] Used with permission.)

---

3   The problems of entering Amsterdam by sea were detailed for me by Mirjam Jonkman, a sailor and Amsterdammer, living in Darwin.

With his local knowledge of winds and tides, Seba could judge his arrival at the *kroeg* with precision.

In the Dutch Golden Age of the 17th century the WIC and its larger and older companion the Vereenigde Oostindische Compagnie (VOC) or East India Company, had brought great wealth to the United Netherlands. By 1732 they were both past their prime. At the outset, the charter given to the Dutch merchants who formed the WIC was a gift of commerce. They were granted a charter by the seven States-General of the United Netherlands in 1621 to trade the west coast of Africa, the east coast of America and the islands of the Caribbean, and to protect the inhabitants of these territories from piracy and extortion. Other merchants found trading in these territories would forfeit their ships and goods.[4] Far from protecting the inhabitants of the Americas, the charter granted the WIC institutionalised piracy and a market reach in slaves, sugar and coffee, enforceable by might. WIC ships worked the triangular route. From the Netherlands they sailed to Africa's Gold Coast where they captured slaves, carried them to South America to work Dutch sugar plantations, then returned to the Netherlands with muscovado raw unrefined sugar – and fortunes raided from other hapless merchantmen.

Sugar was abundant in Holland in the mid-17th century, from where sugar was supplied to much of Europe. More than 50 sugar refineries operated in Amsterdam, providing affordable sugar to satisfy the Dutch *zoetigheid*, or craving for sweetness. *Speculaajes* with exotic ginger and cinnamon, *poffertjes*, waffles and pancakes dusted with sugar all created dental havoc. The Dutch artist Rembrandt, like many of his portrait subjects, suffered the torments of a 'dental cripple'.[5]

In the warmth of the *kroeg* the bosun gave the apothecary the litany of a disastrous voyage to South America. The ship's company had been so diminished by disease and raiding privateers that it was truly a skeleton crew that sailed the vessel with its cargo of muscovado back to Holland. The apothecary saw the telltale signs: the lank hair of malnutrition, a delusional drawn face of scurvy, shivers of residual malaria, and the bug-eyes and bloody lips of dysentery. It was a good time to trade.

---

4   Charter of the Dutch West India Company, 1621.
5   Schama 1997, p. 165.

2 The apothecary's toad

The bosun signalled to a twitchy ferret of a sailor to get the first sack. Earthenware jugs and wooden platters were cleared. As though readying himself for a feast, the bosun spread his arms, broadened his chest and announced to the apothecary that he was about to witness one of the wonders of the Indies. The ferret, drawn erect by the gravity of his role and rising above his threadbare attire ceremoniously placed one of the full sacks on the table then quickly sat, conflicted by the ocean swell of ale and the new-found stability of land. Maintaining the occasion, the bosun unstrung the neck of the sack to reveal parcels of roughly tied sailcloth. Expansively, he unwrapped one of the larger parcels and a gasp of pleasure ran through the gaggle now pressed around the table. A conical shell more than a hands-breadth long emerged, a spectacle enhanced by the rags in which it lay. It flared outwards to produce a lip, a lustrous pink inside, and on the outside, as if etched by a journeyman engraver, a delicate pattern spiralled around the shell to disappear through an aperture large enough for a man's hand to follow. The shell was royally crowned with extraordinary outgrowths decreasing in size as they spiralled to the apex. It was the queen conch, the *caracol rosa*. In the apothecary's Latin patois it was known as *Buccinum ampullaceum striatum, clavicula miricata, apertura leviter purpurascente*.[6] Now classified among strombidae, this conch has the scientific name *Lobatus gigas*.

The bosun knew his market. Conch mania[7] was a fellow traveller with tulip mania and was endemic in the Netherlands. A century earlier, in the 1630s, otherwise careful and conservative Dutchmen lost fortunes on tulip bulbs when the speculative bubble collapsed. But at its peak, besotted owners had each new gloriously coloured bloom engraved on vellum and often in still life together with the most exotic of shells, part insurance and part statement of worth.[8] By the 1730s the Dutch obsession with tulip and conch was more chronic than acute but the bosun knew he would still find ready buyers.

Albertus Seba, like other apothecaries across Europe, had a fashionable *kunstkammer*, or *wunderkammer* – a room in his house holding animals preserved in glass jars, oddities of nature, exotic plants, but-

---

6 Barber 1980, p. 49.
7 Müsch et al. 2001.
8 Goldgar 2007, p. 98.

terflies and insects, shells, corals, rocks, gems and fossils, all arranged for maximum effect. Seba's own collection was exhibited so as to both organise and explain nature.[9] And the stock for Seba's collection, like many others, came from the trade in exotica through the ports of the Netherlands, stimulated by the colonial reach of the VOC and the WIC, and by its entrepreneurial sailors.

In 1732, Seba was building up his display. Fifteen years earlier he had sold his collection in its entirety to Tsar Peter the Great of Russia. The 72 drawers of shells, 32 drawers of around 1,000 insects, and 400 jars of preserved animals so impressed the Tsar that he purchased the collection for 15,000 guilders (around €150,000 today), and shipped it back to St Petersburg.[10] Seba's new *kunstkammer* was to be larger and more comprehensive than the last. He selected the conch and some of the finer shells on offer and his assistant placed them to one side.

The bosun now turned to the damp sack, a special treat for the apothecary. The lump in the sack became animated as the ferrety sailor picked it up from the floor. He carried it at arm's length and with a sideways gait as if to distance it further from his person. He placed it on the table then stood, arms akimbo, pleased with himself, anticipating the reveal. The bosun once more made his pitch to the apothecary – wondrous creatures from the Caribbean, monsters of the natural world, the *aguaquaquan*.

He rolled down the neck of the sack.

A serving wench screamed, dropped her jug and ran. Onlookers stepped back, signed the cross, emitted a low grumble of awe and apprehension.

Two giant warty brown toads squatted in the corner of wet sailcloth blinking their yellow eyes in the dim light. The supernatural on display were no ordinary creatures, quite at home at a witches' Sabbath.

The bosun hushed the throng, calmed the superstitious, continued his sales pitch. He informed the apothecary of the wondrous and curative properties of the creatures before them, of the potent poisons in the sacs on their shoulders, how they devoured all manner of creatures inedible to mankind. They were given life by the salty mudflats of Suriname where they swarmed in great numbers. On the homeward voyage

---

9 Goldgar 2007, Chapter 2.
10 Müsch et al. 2001.

2 The apothecary's toad

they had fed on cockroaches, mice, baby rats, slivers of salted pork rind, and a pair of emerald green lizards from the jungles of Suriname that the ferret had unwittingly placed in the sack with them.

Albertus Seba struggled to hide his enthusiasm for the two mythical creatures. He was grateful for the commotion. Toads were among his *materia medica*, regulated in by-laws since Nicolaes Tulp's *Pharmacopoea Amstelredamensis* of 1636. Powdered toad, taken internally, would reduce fluid retention – oedema or dropsy. Whole dried toads and even live toads, placed over the kidneys, would reduce swelling. Placed on tumours or on the buboes of plague victims they would draw out the poison, 'for thus evil takes away evil'.[11] And a powder of burnt toad, mixed with trisulphate of arsenic, applied to the breast would cure breast cancer – unless the cure killed first. A frog or toad, live or dead, could be bound to a wound as a frog plaster to improve healing.[12] Wearing parts of toads in a bag around the neck was also prescribed to stop bleeding and ward off plague.[13] A cloth soaked in frog or toad spawn – *spermatis ranarum* – dried and then applied to the wound would stop bleeding,[14] as would a frog plaster made from the washings from a number of live toads – *ranas viventes vino lotas dudecim*.[15] And these giant toads had poison sacs – glands full of powerful venom. The giant *ranarum* squatting on the table would supply many potions and their exotic provenance would demand a premium among the apothecary's customers.

The pharmaceutical basis of Seba's remedies was magainin and caerulein, the antibiotic and anti-viral peptide and alkaloid compounds in frog and toad skins. The latter has an analgesic property several thousand times more potent than morphine. And toad skins contain anti-tumour agents, steroids, natural adhesives, and repel parasites and predators including snakes, rats and mosquitoes.[16] Seba's remedies had a scientific pedigree.

---

11   Leeser 1959.
12   Allen 1995.
13   Black 1883, p. 61.
14   Wannan 1970.
15   Koning 1961.
16   Tyler et al. 2007.

In Seba's medicine cabinet, dried, powdered toad would sit alongside roots of valerian, ginger and mandrake; next to lignum vitae, Jesuit bark and cinnamon, leaves of marjoram, oak and oregano, flowers of chamomile, poppy and borage, seeds of cannabis, cardamom and nigella. There would be fruits of hops, figs and tamarind, resins of aloe, gum arabic and opium. And from the ocean, exotic sea sponge, coral, whale sperm and ambergris or whales' vomit. And, even more curiously, the brain of a person who had died violently.[17]

The bosun sealed the deal with a firm hand. Less coin than he hoped for the conch and shells but more guilders than he dreamed of for the toads. He had no other immediate buyers for these ugly creatures. And there were Seba's welcome potions, nostrums, restoratives and remedies included in the deal. Seba prescribed Jesuit bark containing quinine from the *Cinchona* tree for those still shivering with malaria, a tonic of ginger and chamomile with a drop of belladonna for the dysenteric, and a tincture of opium for those demented by scurvy.

Over winter in Amsterdam, Seba's toads became lethargic and lost condition. The two large toads were likely both females and could not mate so he cut short their lives. He dessicated and powdered one and pickled the other in wine vinegar, labelling the jar *Rana marina maxima*, the greatest marine toad.[18] He added the jar to his *kunstkammer*. The bosun would bring him more specimens on the next voyage.

Two years later, in 1734, Seba set about publishing drawings of his collection, a task requiring 446 copperplate engravings in four volumes. Two volumes were published in 1735,[19] one edition in Latin and Dutch[20] and another edition in Latin and French. It became known commonly as Seba's *Thesaurus*.

Seba's *Thesaurus* was not the only scientific publication to benefit from the sugar and slave trade. The WIC helped to popularise the fact that caterpillars, far from being spontaneously generated from mud as

---

17 Koning 1961.
18 Müsch et al. 2001, figure 1, plate 76, folio 1.
19 Müsch et al. 2001, figure 1, plate 76, folio 1.
20 *Locupletissimi rerum naturalium thesauri accurata descriptio – Naaukeurige beschryving van het schatryke kabinet der voornaamste seldzaamheden der natuur* (Accurate description of the very rich thesaurus of the principal and rarest natural objects).

was still widely believed, were a stage in the metamorphosis to butterflies and moths. It was the result of patient observation by artist Anna Maria Sibylla Merian. In 1699, 52 year old Merian followed her eldest daughter when she married a WIC officer who was sent to Suriname to manage a sugar plantation. Her paintings made on the trip, published as *Metamorphosis Insectorum Surinamensium* in Amsterdam in 1705, showed the life cycle of caterpillars, chrysalis and butterflies and their host plants in life-like detail.[21] Merian's paintings captured both the subject creature and key elements of its environment, such as a plant that it fed on. This representation influenced the design, presentation and content of Seba's *Thesaurus*. Indeed, Seba 'borrowed' some of Merian's drawings for his own book.

As Seba glowed in the acclaim reflected from the publication of his first two volumes, he received a visitor, Carolus Linnaeus, a fellow apothecary from Sweden. Linnaeus presented himself in formal traditional Sami robes and bore the good wishes of Herman Boerhaave, Professor of both Botany and Medicine at Leiden and a good friend of Seba.

It was the first of several meetings in Amsterdam between two men whose names were later to be chiselled into the foundation stones of science.

Carolus Linnaeus aspired to membership of Amsterdam's fashionable scientific community. In 1735, at age 28, this son of a Swedish curate had gone to Amsterdam, the centre of European science, to get the all-important title of Doctor. A title that would assure an improvement in both his apothecary's clientele and his fees. It took just one week to purchase a medical degree from the University at Harderwijk with a thesis he had prepared earlier in Sweden. But his private passion was his formative systematic classification of plant and animal kingdoms. Classification was important to apothecaries because different common names for plants and animals collected from different provenances to make medicines caused confusion in the profession. Mistakes over common names could prove fatal to customers. To solve this confusion Linnaeus had devised a systematic classification based on a hierarchy of class, order, genus and species. But he lacked a sponsor to

---

21   Merian 1705. p. 192.

publish it. Penniless, he needed both an income and an entrée into Amsterdam's society to find a patron to publish his *Systema Naturae*.

Linnaeus gained employment in the Amsterdam Botanic Gardens under Johannes Burman, compiling the *Flora Zeylandica* – the flora of Ceylon – from specimens collected by VOC merchantmen. There he was introduced to Professor Herman Boerhaave, an intellectual giant of the sciences in the Netherlands and Fellow of both the French Academy of Sciences and the Royal Society of London. He was Linnaeus's passport to Amsterdam's society and an introduction to three people who proved very important to the rising scientist: benefactor and financier George Clifford, wealthy doctor Johan Gronvius and the apothecary Albertus Seba.

In 1735, Gronvius sponsored the publication of the first edition of Linnaeus's *Systema Naturae*. It was just 12 pages long with the whole of the animal kingdom laid out across a two-page spread. It listed 549 species, more than half of which were referenced to Seba's collection. Seba witnessed the publication but died in 1736 before the second two volumes of his own *Thesaurus* were ready. The final volume of his *Thesaurus* was published in 1765, nearly 30 years after his death.

Carolus Linnaeus returned to Sweden in 1738. When he revisited Seba's *Thesaurus* for the many revisions of his *Systema Naturae*, he would have been reminded of his friend and colleague from the portrait in the first volume. On the frontispiece is Seba, bewigged and robed in his working coat, smiling at the reader, holding a snake preserved in a jar and pointing at the open pages of the *Thesaurus* on a table cluttered with shells and coral. A shelved cabinet behind him carries glass jars containing preserved creatures, and on the top of the cabinet sits a branched coral flanked by two conch shells. It is a portrait of a man happy in his *kunstkammer*.

Seba also acknowledged the sailors and the trade of the VOC and WIC in providing the oddities for his *kunstkammer*. In another frontispiece, an allegorical scene of *Industria*, a group of characters clad in a Greek classical style hold a conversation around the four volumes of Seba's *Thesaurus*. 'Industry' in female form discourses with 'Commerce' and 'Science' watched over by 'Truth' while old man 'Time' listens, leaning on his scythe. In the near background three indigenous people bearing goods represent the Americas, Africa and Asia. And on the ocean in the far background, square-rigged three-masted ships sail to

2 The apothecary's toad

Figure 2.2 Toad in Seba's Thesaurus referred to by Carolus Linnaeus as Rana marina (Tomus 1 Tabula 76). (Den Haag, Koninklijke Bibliotheek. [Locupletissimi rerum naturalium thesauri accurata descriptio Naaukeurige beschryving van het schatryke kabinet der voornaamste seldzaamheden der natuur/Albertus Seba.] Used with permission.)

far off colonies bearing the distinctive horizontal bands of red white and blue of the pennants shared by both the VOC and the WIC.

Linnaeus's 10th edition of *Systema Naturae*, published in 1758, is considered the starting point of the modern Linnaean systematic classification because it was this edition that, among other things, correctly placed whales with mammals rather than fish.[22] In this edition Linnaeus named Seba's conch shell *Strombus gigas* and referred to six drawings of anurans from Seba's collection which he renamed as five species; one of these was *Rana marina* – Seba's original *Rana marina maxima*.[23] Linnaeus described *Rana marina* simply as having a lumpy back and knotty hindquarters, four toes on its front feet, and five par-

---

22   Linnaeus 1758, p. 194.

tially webbed toes on its back feet: '*Rana scapulis gibbosis, clunibus nodosis. Palmae tetradactylae fissae; plantae pentadactylae subfissilae*'.[24] In 1799, Strasbourg scholar and naturalist Johann Gottlob Theaenus Schneider revised the toad's name to *Bufo marinus*.

Linnaeus's *Systema Naturae* came to form the essential scaffold for classification in the biological sciences and the last two components of his classification, genus and species, became the universal binomial naming system scientists use today. Its germ was a Swedish apothecary seeking clarification for his professional world. Its birth in Amsterdam came with the help of Seba's *kunstkammer*, the financial generosity of Johan Gronvius, the sailors of the WIC and VOC and, in turn, the Dutch trade in sugar and spices. Seba's *Rana marina maxima*, carried from coastal mudflats of Suriname with a cargo of muscovado, ended up preserved *in vitro*, in the copperplate engravings of his *Thesaurus* and in Linnaeus's *Systema Naturae*.

The sugar industry and the apothecary's toad, *aguaquaquan, Bufo marinus*, the cane toad, have been intimately linked ever since.

The toads did not do well in Amsterdam because they were South American poikilotherms – cold blooded amphibians. Unlike mammals they could not increase their body temperature by metabolic activity and had to rely on ambient heat, a quality lacking in wintry Amsterdam. If the apothecary's toad had evolved in a cool temperate environment high in the Andes, it may well have become a resident of both Europe and Asia, spreading east and west of the Urals over the next 300 years. It would have required only a breeding pair exchanged for just a few coins and some palliative nostrums.

---

23   In the Taschen edition of Seba's *Thesaurus* (Müsch et al. 2001) *Bufo marinus* is depicted with distinctive warty skin and parotoid glands in figure 1 in plate 76, folio 1. Other engravings (figures 1 and 2 in plate 73, folio 1) also have a modern annotation as *Bufo marinus*, but neither has distinctive parotoid glands and are referred to by Linnaeus in the 1758 10th edition of *Systema Naturae* (p. 213) as *Rana arboreta*.
24   Linnaeus 1758, p. 211.

2 The apothecary's toad

Figure 2.3 Industria. (Den Haag, Koninklijke Bibliotheek. [Locupletissimi rerum naturalium thesauri accurata descriptio Naaukeurige beschryving van het schatryke kabinet der voornaamste seldzaamheden der natuur/Albertus Seba.] Used with permission.)

3
The sweet grass

In 1493, almost 40 years before Martim Afonso de Sousa introduced sugar cane, the sweet grass, to the mainland of South America, Christoforo Colombo introduced sugar to the island of Hispaniola (now the Dominican Republic and Haiti). This Genoese sailor was sponsored by King Ferdinand and Queen Isabella of Spain, and these early plantings created a very profitable plantation industry supported by slave labour abducted from Africa. But soon, profits had to be shared with pests as sugar cane estates were planted throughout the Caribbean, into the Pacific and on to Australia. Some pests, like rats, accompanied man; others were distributed together with growing stock of sugar cane; yet more pests, local fauna, presented with an abundance of sugar on their doorsteps took full advantage. The rise in pests of sugar cane and the absence of effective pest controls gave birth to biological control. To protect the sweet grass, sugar cane scientists would release wasps, flies, beetles and the mongoose to combat pests in a biological arms race that would eventually include the cane toad.

In the 17th century, Spain and Portugal's sugar trades attracted attention from European powers. On the Monopoly board of the Caribbean, European powers rolled the dice, then argued about whose turn it was, stole pieces, squabbled over property, ignored the community chest and refused to go to jail. A snapshot of the game in the early 18th century, around the time Albertus Seba purchased his Suriname toads from WIC sailors in Amsterdam, would show the players on the board. The Dutch had sugar estates in Suriname on the north coast of South America; Spain held sugar territories in Cuba, Hispaniola and Puerto Rico; Portugal retained its hold on sugar estates in Brazil; the French were in French Guiana neighbouring Suriname, the Caribbean island of Martinique and on what would later become Louisiana on the North American mainland; Great Britain's sugar estates were in British Guiana neighbouring Suriname and on Barbados, Jamaica, Trinidad, Antigua and other small Caribbean islands. It was a complex and constantly changing field of play with pirates playing in a breakaway league on the high seas.

Good profits in European markets made the sugar industry a lure for investors.[1] A walk through a stand of Caribbean sugar cane, hemmed in by stalks, a canopy of leaves closing overhead, an airless mass of green, would confirm the abundance of production, the simplicity of growing cane, would impress the visitor and convince him to part with his money. But without the technology to extract cane juice and manufacture marketable sugar, the investment in growing cane was worthless.

Today, sugar cane is grown from 12 months to two years before being harvested when concentrations of sugar in the juice are highest. Cane stems are crushed between steel rollers, juice collected, heated, alkali added to precipitate impurities, then filtered, boiled in evaporators and vacuum pans to reduce volume and initiate crystallisation. A centrifuge separates molasses from raw sugar that may be refined further using acids, preservatives, flocculants, surfactants and bleaches – this is modern sugar chemistry.

In the 17th century, steps were missed or misjudged. Cane harvested too early or late or left too long before crushing yielded little or no juice. Omitting alkali created a dirty, sticky mess. Volume reduction and crystallisation was haphazard, and draining-off molasses by gravity was unpredictable. A brown, sticky muscovado was exported to Europe. Once there, molasses was again washed from sugar in a conical mould and the sugar then left to cure. The retail product was a solid, conical 'sugar loaf' weighing between 5 to 35 pounds (2.3 to 16 kg).

---

1   Molen 1971.

3 The sweet grass

When visiting the Caribbean in the 1640s looking for an investment opportunity, Englishman Richard Ligon noted that on Barbados 'the Sugars they made were but bare Muscavadoes, and few of them Merchantable commodities; so moist, and full of molasses, and so ill cured, as they were hardly worth the bringing home for England'.[2] But help arrived. On the South American mainland, Portuguese took back sugar estates from the grasp of the Dutch, who fled to Barbados where their expertise helped solve technical problems in the production of sugar.

By the 18th century, Barbados was the main supplier of muscovado to England but the sweet grass was beset by pests. Rats were everywhere in the Caribbean. As early as 1688 on Jamaica Sir Hans Sloane remarked that rats were 'all over the Island, both in Houses and Lands where they destroy the Sugar Canes, by eating some and barking others. They are taken and swallow'd whole by the Snakes, for which good Service these last are not molested'.[3] Rats reduced the yield of the cane harvest and, in some years, large sugar ants, *Formica omnivora*, descended in plagues destroying cane crops and livestock. In 1760 these ants caused such devastation on Barbados that 'it was deliberated whether that island, formerly so flourishing, should not be deserted'.[4] Caterpillars (*Diatraea saccharalis*) and palm weevils (*Rhynchophorus palmarum*) bored into the tips and stems of cane, cane flies (*Saccharosydne saccharivora*), aphids and ants made it worthless for processing and the cane would then 'not only yield nothing at the mill, but communicate a dark colour and bad quality to the syrup'.[5] And, to top it off, hurricanes destroyed both crops and sugar mills.

Barbados, the most easterly island of the West Indies, is the first landfall on a voyage from Europe. Its coral geology forms a gentle relief of flat terraces rising stepwise to 340 metres (1,120 ft) above sea level, and its fauna, like many islands in the Caribbean, is poor in biodiversity and low in number because of isolation from neighbouring continents. Man introduced most pests to the island. Rats jumped ship. Forest soil and leaf litter, imported from the mainland to improve soil fertility,

---

2 Ligon 1657.
3 Sloane 1725, p. 330.
4 Schomburgk 1848.
5 Schomburgk 1848, p. 645.

contained destructive sugar ants and termites.[6] And with all available timber on Barbados burnt in the sugar mills, wood was also imported, together with hidden pests.

There was no effective pest control. Islanders combated ants with fire, tar, arsenic or corrosive sublimate – mercuric chloride – also used to treat syphilis. And the Catholic population of Hispaniola, invoking pest control from on high, made a procession in honour of St Saturnin to save their crops. But whatever the method, the outcome was the same. With nothing left to eat, ants eventually disappeared[7] leaving the inhabitants the job of starting their cane plantations anew.

Rats chewed the stalks of sugar cane, feeding off the juice. They carried bubonic plague and multiplied in great numbers under the dense cover of the cane crop. In 1832, the giant toad, *Bufo marinus*, was introduced to Barbados from Demerara in British Guiana, to eat rats. There was proof of their efficacy – albeit third-hand. The historian, Schomburgk, related that 'a gentleman whose veracity I have no reason to doubt has assured me he witnessed a combat between one of these toads and a rat, in which the toad succeeded in driving the rat from the field'. Toads liked Barbados and by 1847 they were 'to be met with in as large numbers as in Demerara [on the South American mainland]'.[8] Toads were also taken from Barbados to Jamaica. Jamaicans took a dislike to them and killed them at every opportunity,[9] but toads bred in such numbers that eradication proved impossible.

Toads had a pedigree for pest control. In France, for much of the 19th century, European toads (*Bufo bufo*) could be purchased in toad markets: 'One of Paris strangest industries is without doubt the toad markets ... The clients of toad merchants are gardeners from Paris surroundings who, lacking time to do otherwise during the week, protect their lettuces and other plants with these efficient amphibians, as well as, to the ladies dismay, card readers and fortune tellers who also go through a large number of toads [searching their innards for signs].'[10] Toads were sold in Paris markets in wasteland near Feuillan-

---

6   Schomburgk 1848, p. 166.
7   Schomburgk 1848, p. 640.
8   Schomburgk 1848, p. 679.
9   Waite 1901.
10 Kuble 1879.

tines Street[11] and in Le Jardin des Plantes[12] at two francs fifty centimes per dozen.[13] Among local buyers were Englishmen who crossed the Channel to collect toads. Toads were sold in England for around £13 per hundred.[14] Similarly, across the Atlantic in 1870 in Hartford, Connecticut, Charles Dudley Warner considered the toad 'the most useful animal in the garden [and] … an animal without which no garden would be complete' although it would not rival 'the completeness of the Paris "Jardin des Plantes" '.[15]

Naturally, French planters on the island of Martinique, to the northwest of Barbados, also thought to employ toads to control pests in sugar cane, especially since the toads on the mainland were so large and had such big appetites. They introduced giant toads, *Bufo marinus*, to Martinique from Cayenne in French Guiana on the north coast of South America.[16] Some of these toads were also shipped from Martinique to Barbados prior to 1844.[17]

On Barbados, toads were conspicuous because they were the only amphibians on the island. But toads could not solve the sugar industry's problems of declining production. After 200 years of continuous cane cropping, soils on Barbados were depleted of nutrients and soil structure. The fertility of once virgin soil was gone and crops were stressed and prone to attack from diseases and pests. The solution to the problem was to expand sugar plantations on British Guiana on the South American mainland – to establish new plantations on fertile virgin soil freshly cleared of forest cover. By 1882 sugar exports from these new continental plantations had grown to twice those of Barbados and Jamaica combined.[18]

In the 19th century, new growers entered the sugar market. Sugar production increased in Spanish Cuba, and in Louisiana which the United States of America purchased from France in 1803. Sugar plant-

---

11   Kuble 1879.
12   Autour de la "Mouffe", le turbulent Fabourg St Marcel, Paris Révolutionnaire (n.d.).
13   Riley 1869.
14   Kirkland 1904, p. 14.
15   Warner 2006(1870).
16   Waite 1901.
17   Easteal 1981.
18   Williams 1970, p. 366.

Figure 3.1 Toad market in Paris 1879. (Le Journal illustré, Dimanche 7 Septembre 1879.)

ations were also being developed in the Hawaiian Islands in the middle of the Pacific Ocean, and on the Pacific coast of Queensland, Australia.

As with Barbados, biodiversity on Hawaiian Islands was simple, the islands being the result of mid-ocean volcanic eruptions. As it emerged, each initially pristine island along the Hawaiian chain was colonised by plants and animals that reached it by wind, water, or wings. Prior to human settlement, there were about 653 species of plants and 2,740

3 The sweet grass

species of insects endemic to the islands.[19] Around 2,500 years ago, migrating Polynesians introduced taro, sweet potato and sugar cane to Hawai'i. Their sugar cane came from the plant's origins, the Pacific island of New Guinea, where some 37 species of *Saccharum* range from the very sweet 'noble cane', *Saccharum officinarum*, to sugarless grass *Saccharum spontaneum*.

The rulers of Hawai'i gained revenue from allowing foreigners to exploit natural resources like sandalwood, seal fur and whales. Sugar cane was a subsistence crop. This changed on 6 May 1825 when the 46 gun British man-of-war HMS *Blonde* dropped anchor in Honolulu. The agent of change was an Englishman, John Wilkinson, a sugar planter from the West Indies. Technical expertise in sugar manufacture arrived in Hawai'i.

It was a sad occasion for the islanders. Below decks lay the bodies of King Kamehameha II and his wife Queen Kamamalu in triple-layered caskets of lead, mahogany and oak. In London the year before to visit the King of England, all Hawaiians on the party had caught measles; they had no inherited immunity. First the Queen died and then the distraught King.[20] King George IV commissioned HMS *Blonde* for the funeral voyage. Five days after the arrival of the *Blonde* in Honolulu the caskets were carried ashore. Befitting his orders from the Admiralty, the *Blonde's* captain, Lord Byron, cousin of the famous poet, arranged a funeral procession from the warship to Honolulu's thatched church. The procession was led by 12 Hawaiian warrior chiefs in feather cloaks and helmets along a route lined by Hawaiian guards resplendent in fragments of Russian uniforms. Royal Marines followed in red tailcoats, black shakos and polished brasses, then the ship's band, the chaplain and surgeon in full-dress blue tailcoats, trailed by two dowdy American missionaries. Only then did the caskets appear, each draped in black and drawn by 40 chiefs, followed by foreigners including Wilkinson, captains of the ships in port and 100 sailors from the *Blonde* in smart white uniforms.[21] It was the British navy at its imperial best and Honolulu had never seen its like before.

---

19   Perkins 1913.
20   Kuykendall 1938, p. 119.
21   Joesting 1972, p. 79.

In the funeral procession were the key ingredients to lift the sugar industry from a subsistence crop: Hawaiian chiefs who controlled land, missionaries who advised and increasingly directed the chiefs, American capital, and sugar manufacturing expertise from the West Indies.

John Wilkinson established seven acres of sugar cane in the upper part of the Manoa valley behind Honolulu and made some sugar. Lack of both equipment and funds compounded Wilkinson's poor health and he died after only 18 months on Oahu. The Governor of Oahu, Governor Boki, had invited Wilkinson to Honolulu; he now found four new foreigners to invest in his enterprise and converted the nascent sugar mill into a more profitable rum distillery. But the new Queen Regent, Kaahumanu, recently converted to Christianity, placed a *kapu*, or *tabu*, on rum. American missionaries owned the only ox carts in Honolulu and used their monopoly to stop cane being carried to the mill. Boki's enterprise foundered once again.[22]

Missionaries helped kill the rum venture but they appreciated the benefits for Hawaiians of organised enterprise. In 1829, a missionary on Hilo announced the production of a year's supply of sugar and molasses from a hand-powered wooden mill. Innovation within their own ranks led missionaries to encourage 'the growth of cotton, coffee, sugar cane etc., that the people may have more business on their hands and increase their temporal comforts'.[23] With the support of Hawaiian chiefs and missionaries, in 1835 William Ladd of Ladd and Co established the first permanent sugar cane plantation in the Hawaiian Islands at Koloa Plantation on the island of Kauai. The beginning was just 25 acres and a 'rude affair' of a wooden mill.[24] Soon small plantations and mills sprouted up on the islands of Maui, Oahu and Kauai.

By 1846, 22 sugar mills operated on the Hawaiian Islands producing a modest export trade of around 500,000 pounds (226 tonnes) per annum.[25] But two events on the North American mainland were about to stimulate the sugar industry in Hawai'i.

First was the discovery of gold in California.

---

22   Kuykendall 1938, p. 173.
23   Kuykendall 1938, p. 174.
24   Kuykendall 1938, p. 175.
25   Kuykendall 1938, p. 316.

3 The sweet grass

On 24 January 1848, James W Marshall was overseeing the construction of a water race to power a sawmill on the south fork of the American River at Culuma[26] in California's Sierra Mountains. Thirty-three year old Marshall, carpenter and farmer, was 'a cool enough man except where his pet lunacy [spiritualism] was touched'.[27] He and his construction crew were from the recently disbanded Mormon Battalion, raised for the Mexican–American War. On that Monday morning, Marshall inspected the construction of the water race; in the evening, he announced to his workmates: 'Boys, I believe I have found a gold mine'.[28] News spread. The Californian gold rush was on. San Francisco's population swelled from 2,000 souls in February 1848 to 20,000 by year's end.[29]

News reached Honolulu in June 1848 on the schooner *Louise* carrying two pounds of gold as tangible proof. Expatriate Europeans and native Hawaiians joined the rush. By August, 1,000 pickaxes had been exported from Honolulu together with men to wield them.[30] It was a welcome change from the whaling industry, considered 'a fluctuating and precarious prosperity not rooted in the soil of the kingdom'.[31] Thanks to the gold rush, Hawai'i experienced a 'potato boom', exporting more than 50,000 barrels of Irish potatoes and nearly 10,000 barrels of sweet potatoes to the gold fields in 1850. The potato boom was over just a year later due to cheaper production locally in California. By contrast, sugar and coffee was not easily substituted in the gold fields, and demand for these crops grew despite the economic depression of 1851. It stimulated investment in new plantations of sugar and coffee in Hawai'i and in new technology like centrifugal machines that, within minutes, separated sugar from molasses and increased the quality and yield of sugar. By 1861, 2.5 million pounds (1,200 tonnes) of sugar were leaving Hawai'i's shores; a five-fold increase in sugar exports in 15 years.[32]

---

26 Later called Coloma.
27 Bancroft 1888, p. 31.
28 Bancroft 1888, p. 31.
29 Bancroft 1888, p. 31.
30 Bancroft 1888, p. 31.
31 Kuykendall 1938, p. 310.
32 Kuykendall 1938, p. 315.

The second stimulus to the Hawaiian sugar industry was the American Civil War.

On Sunday 14 April 1861, Major Robert Anderson, Kentuckian, 56 year old professional soldier and artillery expert, stood to attention among the burning remains of Fort Sumter, Charleston, South Carolina. He saluted the flag of the Union and surrendered the fort to Confederate Brigadier General Beauregard. What was left of his company formed up on the parade ground and 'marched out upon the wharf, with drum and fife playing Yankee Doodle'.[33] President Abraham Lincoln declared war on the Gulf States. New York's broadsheet, *Harper's Weekly*, announced 'The die is now cast, and men must take their sides, and hold to them'.[34]

President Lincoln declared a blockade of Confederate ports on the Atlantic and Gulf coast: New Orleans, Mobile, Richmond, Charleston, Savannah and Wilmington. Exports of sugar and cotton all but ceased. Before the blockade, Louisiana harvested around 230 million pounds (104,000 tonnes) of sugar, but after the blockade production was less than 20 million pounds (9,000 tonnes). The price of Louisiana sugar went from around $70/hogshead before the blockade to $157/hogshead afterwards.[35]

Blockade runners, manned by British Royal Navy officers on leave, carried insignificant cargoes, and the blockade shifted the focus of the sugar trade to the cane fields of Hawai'i. Sugar exports from Hawai'i went from 2.5 million pounds in 1861, the year before the blockade, to 15 million pounds in 1865 after the end of the war,[36] a six-fold increase in just four years. It was spectacular growth and the American Civil War helped embed the sugar industry in the economy of the Hawaiian Islands.[37]

For the most part, Hawai'i was free of pests and diseases. But good economic times brought more trading vessels with goats, cows, horses, dogs, cats and poultry as well as rats, mosquitoes, cockroaches, termites, ants, bedbugs, weevils and moths. And visitors brought new food

---

33   *Harper's Weekly*, 27 April 1861.
34   *Harper's Weekly*, 27 April 1861.
35   Anon. 1882.
36   Kuykendall 1953, p. 141.
37   Kuykendall 1953, p. 140.

plants along with soil and new pests. In 1865, sugar cane in the Lahaina area was damaged by the sugar cane weevil borer *Rhabdoscelus obscurus*. The weevil was probably introduced to Hawai'i 11 years earlier, brought to Hawai'i from Tahiti by Captain Edwards of the whaling ship *George Washington*. Edwards gave sugar cane to Consul Chase of Lahaina, Maui, who planted it in his garden. Chase distributed cuttings to his friends who in turn gave cuttings to their friends. This generosity ensured that adult weevils and eggs were distributed throughout the Hawaiian Islands and that the borer became a serious economic pest of sugar cane.[38] Lack of plant quarantine, ignorance of pests and diseases and the ease with which they were spread meant the abundance of the Hawaiian sugar industry was now shared with pests.

Each Hawaiian island was a Petri dish. Remote in a blue ocean, each new wave of visitors inoculated the fertile volcanic soil. Water and sunlight helped new life forms flourish in the tropical incubator. Introduced life forms took over each dish. On the Hawaiian Islands, as in the Caribbean, the simplified structure of the islands' biodiversity meant pest populations could achieve destructive proportions and there was little that could be done to control them.

In the 19th century, Hawai'i was not the only country to seek to benefit from the hiatus in sugar, cotton and tobacco markets created by the American Civil War. On the other side of the Pacific Ocean, in Queensland in northern Australia, entrepreneurs were also keen to take advantage. Courtesy of the Civil War, the sugar industry also became established in Australia. And like the West Indies and Hawai'i, in time they too would be beset by pests.

---

38   Pemberton 1964.

4
Queensland sugar – Hope and Whish

April 1862, Brisbane, Queensland, Australia. Three men met on the banks of the Brisbane River in the Botanic Gardens just on the fringe of town. Brisbane, the capital of the fledgling Colony of Queensland, was the former British convict settlement of Moreton Bay. The gardens were home to the colony's collection of potential agricultural crops – plantings of wheat, grapes, tobacco, tea, coffee, mango, paw paw, ginger and sugar cane. For the trio, an investor, a botanist and a sugar technician, it was a good venue to discuss market prospects for sugar in Queensland.

Queensland had seceded from the colony of New South Wales just three years earlier, in 1859. At separation, the colony's immigrant white population numbered around 30,000 – aboriginal people were not counted. There were more than four million sheep in the colony and wool accounted for 78% of Queensland's exports.[1] To prosper, the colony needed to attract more settlers and diversify its agriculture. *Pugh's Queensland Almanac, Directory and Law Calendar*, read in detail by new settlers, promised that in Queensland 'most, if not all, the productions of the Indies, South America, and not a few of those of Africa, may be successfully, and therefore profitably, cultivated'.[2] Cotton had proved vulnerable to Queensland's seasons. Sugar cane grew well but,

---

1 Fitzgerald et al. 2009, p. 15; Pugh 1863.
2 Pugh 1863.

as at the start in both Barbados and Hawai'i, there was no resident technical expertise to make crystalline sugar from cane juice.

Further south in New South Wales, attempts to produce sugar in 1823 had failed. Thomas Alison Scott, a Scottish sugar planter from the West Indies, had manufactured sugar but was sacked by the NSW Governor.[3] The colony continued to be supplied with Philippine sugar refined in Sydney.[4] The warmer climate of Moreton Bay and the town of Brisbane was more suited to growing sugar cane than Sydney. In 1849 a crop was grown just north of Brisbane at Eagle Farm but no sugar made from it.[5] In 1850 there was another attempt. A baker boiled juice from sugar cane harvested from Brisbane's Botanic Gardens and produced three kilograms of sugar, but nothing more.[6] Now, in 1862, there had been a more successful attempt. It was the reason for the meeting.

The tallest of the trio was a Scotsman, Captain The Honourable Louis Hope, in town to attend the new Parliament of Queensland. A man of rank, he bore the pedigree and bearing of the Coldstream Guards, the 'Lilywhites'. A silk top hat enhanced his stature. Its swept brim and his groomed sideburns framed a disdainful expression. He held a silver topped cane in the manner of a swagger stick under the right arm of his frock coat. He flourished it at times to emphasise a point. Hope had left England in 1843 aged 26 and still a bachelor. He was headed to Colombia, sailed to Sydney and then on to the Sandwich Islands (Hawai'i) and the Society Islands (Tahiti). He was bewitched by the valley of Hanalei[7] on Kauai where he dreamt of owning land and cattle and 'induced to be a S[andwich] Islander for a few years'.[8] Instead, on Hawai'i on his 27th birthday in 1844, he lamented in his diary 'How much further will my sojourn thereon [this earth] be prolonged and why have I been preserved hitherto while five brothers and two sisters have been called away all of them better and worthier than myself?'[9]

---

3   Bell 1956, p. 23.
4   Philippine sugar was refined by the Australasian Sugar Company in Sydney, later to become the Colonial Sugar Refinery or CSR. Lowndes 1956, p. 13.
5   Bell 1956, p. 23.
6   Fitzgerald 1944.
7   For many, except the songwriter, Hanalei is the home of *Puff the Magic Dragon*; the song recorded in 1963 by Peter, Paul and Mary. It was one of the locations for the 1958 film *South Pacific*.
8   Hope 1844.

The seventh of nine sons, Louis Hope was at the far end of the line for the family title, even with the regular mortality of his siblings. But he was maintained well enough by the family seat, the grand Hopetoun House,[10] called the Scottish Versailles,[11] on the shores of the Firth of Forth outside Edinburgh.

In Hawai'i Hope resolved to return to Australia. He settled in Queensland on a property at Cleveland to the south of Brisbane that he called Ormiston,[12] after his family's agricultural estates in East Lothian, Scotland. There, among other trial crops, he planted 20 acres (eight hectares) of sugar cane. In this he was supported by Queensland's Governor, George Bowen, who declared in 1860 that 'with sufficient capital and under efficient management, the cultivation of the cane for sugar ought to prove one of the most profitable arrangements which offer themselves in Australia'.[13] But Hope had found no one in the colony experienced in the manufacture of sugar.

The shortest of the trio, a stocky Scotsman, Walter Hill, stood erect with the balanced stance of a lifetime out of doors. Eyes shone out from a profusion of agrarian whiskers – a greying hayfield of hair topped snowy hedgerows of eyebrows and sideburns, and an autumnal copse of chest hair escaped the confines of a high collar. His rounded, weathered face was softened by the sunken lips of a mouth untroubled by teeth, but his strong jaw, held at a defiant tilt, defined a man confident of both status and demesne. As Superintendent, the Botanic Gardens were Hill's fiefdom. He was also Queensland's Government Botanist. His was a practical pedigree: apprenticed at age 15 in the gardens of Balloch Castle, Dumbarton, then nurtured in the finest botanical traditions as a propagator in the Royal Botanic Gardens, Edinburgh, and Kew Gardens, London.[14] Now he was lord of nine acres of sloping land and guardian of experimental plantings. In 1862 he distributed 8,000 plants and 700 packets of seed sourced from the Botanic Gardens to prospective growers throughout the colony. Among the

---

9   Hope 1844.
10 A History of the Hope Family, Hopetoun (n.d.).
11 Hopetoun House (n.d.).
12 Ormiston House Friends and Advisers Committee (n.d.).
13 Graves 1993, p. 11.
14 McKinnon 2007.

Figure 4.1 Captain The Hon. Louis Hope. (State Library of Queensland. Negative number 66902.)

plants, '1,600 Tea, 200 Coffee, 30 Cinnamon, 20 Tamarind, 34 Custard Apple, 4 Mango, 3 Alligator Pear, 3 Longan, 160 roots Ginger, and 2,000 Sugar Cane'.[15] Hill was the arbiter of what would grow where.

---

15   Pugh 1863.

4 Queensland sugar – Hope and Whish

Figure 4.2 Walter Hill, Superintendent, Queensland Botanical Gardens. (State Library of Queensland. Negative number 17483.)

John Buhôt, the third much younger man, was lean and slight of build. Incipient whiskers and thin brown hair gave him a boyish look. His agile eyes lacked the hoods and bags of wisdom. He gesticulated, tactlessly punctuating arguments with short stabs of half-chewed sugar cane. He was a sugar maker, recently arrived in Moreton Bay with his new wife from the sugar plantations of Barbados via London.[16] He quickly gained a reputation, reportedly having 'many real friends and admirers but it was no uncommon thing for him to abuse them very roundly'.[17] He was aggressively confident of growing and processing sugar and of making his fortune – from the fortunes of others. Fortune hunters arrived in the colony with each new ship. In the acerbic assessment of the Colony's first Premier – its Colonial Secretary, 28 year old Robert George Wyndham Herbert – they were 'well educated young men, who come here under the general idea that every man may do well, and who are now beggars ... many become drunkards'.[18]

On 25 April 1862 John Buhôt manufactured granulated sugar in Walter Hill's cottage using juice squeezed from cane from the gardens.[19] Legend was he did it 'with three pots and a candle'.[20] Buhôt believed he could garner financial gain from this demonstration of his technical prowess. Hope could provide capital, Hill the plants, and Buhôt technical expertise. And it worked. After the meeting, Louis Hope employed John Buhôt to manage his property at Ormiston. There was newly planted sugar, corn and arrowroot, salt pans for salt production, stone for building, stockyards and milking yards, a barn, slab huts, a small brick house and 'a beautiful site for a house facing the bay and the garden stocked with oranges, vines, bananas and no end of things'.[21]

A few months later, in September 1862, Hope offered to sell Ormiston to a 'new chum', young Englishman, Captain Claudius Buchanan Whish, late of the Kings Light Dragoons, India. He was newly arrived in the colony on the *Young Australia*. Claudius Whish followed his father, General Sir William Whish, into the army but, unlike Hope, had no in-

---

16 Wood 2007.
17 Wood 2007.
18 Fitzgerald et al. 2009, p. 19.
19 Pugh 1863.
20 Whish 1863a.
21 Whish 1862a.

herited fortune. He resigned from the army and migrated to Moreton Bay with his wife Annie, two daughters, two young female servants, two young male apprentices, two dogs, household furniture, a collection of

Figure 4.3 John Buhôt. (State Library of Queensland. Negative number 98180.)

plants and seeds, and life savings of £3,505.[22] This professional soldier planned to make his fortune from farming. Only his abundant optimism exceeded his cloistered naivety. As a former army captain, Whish was feted on arrival in the small society of Brisbane. Within 10 days of stepping ashore, dining with the Queensland Governor, Irish-born Sir George Bowen, and his wife, the Greek-born Contessa Diamantina Roma,[23] Whish was told 'the cultivation of sugar will pay well ... it does not suffer from the changes of climate like the more delicate cotton'.[24]

Whish could not afford to buy Hope's property at Ormiston – at £6,500 the price was twice his savings.[25] He chose instead to settle on land on the Caboolture River, to the north of Brisbane. Reflecting on former employment on half pay as musketry instructor in Dublin, he considered this 'surely not worse than Newbridge [Barracks] for a poor married man!'[26]

Hope sacked the argumentative Buhôt in December 1862. Whish hired him directly. Together, Whish and Buhôt planted a crop of sugar cane which 'in October next should give us 10 acres yielding at least £1,000 worth of sugar'.[27] It was a fanciful estimate. Six months later Whish sacked Buhôt because labourers refused to work under him.

In September 1864, Louis Hope milled the first sugar crop in the colony of Queensland. He won the medal and prize put up by the London Society of Arts.[28] From his eight hectares of cane, using a mill imported from Glasgow, he produced three tons of raw sugar, and three quarters of a ton of molasses. Claudius Whish's crop followed soon after. Building on success, the Queensland Government introduced Sugar and Coffee Regulations to encourage settlers to take up land at a nominal rent provided they grew one of these crops.

---

22   Whish 1863b.
23   The Governor and his wife are commemorated by the town of Bowen in northern Queensland, the Diamantina River and the town of Roma in western Queensland, and the names Bowen and Roma can be found in suburbs, streets and train stations in Brisbane.
24   Whish 1863a.
25   Whish 1862a.
26   Whish 1862b.
27   Whish 1863b.
28   Anon. 1935, p. 406.

4 Queensland sugar – Hope and Whish

Figure 4.4 Captain Claudius Buchanan Whish. (State Library of Queensland. Negative number 69359.)

Clearing of forested coastal land for sugar cane proceeded well in advance of the development of the roads necessary to carry agricultural products to mills and markets. This led to innovation: instead of taking sugar to the mill, the *Walrus*, a small wooden-hulled steam vessel housing the Pioneer Floating Sugar Mill Company, took the mill to the sugar. On board, sugar cane went in one end of the *Walrus* and rum came out the other – bets were made and lost in between. The *Walrus* meandered along the Logan River, south of Brisbane, pleasing cane growers, its crew and the general populace.

In 1864, as Hope produced the first three tons of sugar in Queensland, Hawai'i produced 4,500 tons of sugar. Then in 1865 the American Civil War came to an end. On 9 April, surrounded by Union forces, General Robert E Lee surrendered the Confederate Army of Northern Virginia to Lieutenant General Ulysses S Grant at Appotomax Court House. On 20 August 1866, President Andrew Johnson declared the end of the American Civil War. Americans paused to count losses but in Hawai'i and Queensland investors counted profits. In 1867, just two years after the end of the war, some 10,000 acres (4,000 hectares) of sugar plantations had been established in Hawai'i,[29] and 5,000 acres (2,000 hectares) of coastal Queensland forests had been cleared for sugar cane.[30]

Sugar was not the only new crop in Queensland. There was a purposeful intent by new settlers to populate these new lands with as many exotic but useful plants and animals as they could. Encouragement came from acclimatisation societies, promoted by eccentric men with even more eccentric ideas. Begun in France, acclimatisation was championed in England by Francis Trevelyan Buckland, Assistant Surgeon in the 2nd Life Guards. Buckland was 'four and a half feet in height and rather more in breadth – what he measured round the chest is not known to mortal man'.[31] After a memorable dinner with fellow zoöphagist Professor Richard Owen, superintendent of the Natural History Department of the British Museum, Buckland and Owen formed the Society for the Acclimatisation of Animals, Birds, Fishes, Insects and Vegetables within the United Kingdom.[32] Its President was

---

29   Dorrance 2001, p. 6.
30   Anon. 1930, p. 5.
31   Burgess 1967, p. 59.

4 Queensland sugar – Hope and Whish

the Marquis of Breadalbane who husbanded African eland, Himalayan yak and American bison on his Taymouth estate in Scotland. Following in train, in 1862, Queensland Governor George Bowen established the Queensland Acclimatisation Society 'to obtain from various parts of the world seeds, trees, plants and animals possessing intrinsic value'.[33] In the grounds of Bowen Park, Brisbane, the Society established bananas, cotton, apples, jute, pineapples, grape vines, poppies, wheat, chicory, coffee, mulberries, sorghum, cocoa, tea, and sugar cane.[34] The Society also released fallow deer, axis deer, red deer – a gift from Queen Victoria – rusa deer, angora goats and llamas. Few survived hungry poachers, thereby proving the premise of their introduction. And its members also released skylarks, blackbirds, song thrushes, rooks, hedge sparrows, house sparrows and starlings – most became pests.[35] The Society also championed prickly-pear cactus for prickly cattle fences but it quickly became an uncontrollable pest on grazing land. It confirmed the Society's pedigree for introducing plants and animals that would become pests in their new habitats.

New crops also became the target of indigenous pests. There were rats, both native and introduced, and the new succulent crops, growing unprotected on land cleared from coastal native forest, proved too great a temptation for the larvae of many native beetles and moths that also became pests.

Buhôt was on hand to help. Based on his experience in Barbados, his advice was that 'rats are the greatest pest of the planter. No other vermin ... need be feared'.[36] Buhôt did not mention anything about the use of toads to control rats in the Caribbean. And, thankfully, he had no experience of the Indian mongoose,[37] a cat-sized carnivore introduced in the early 1870s to the West Indies by sugar planters to control rats.[38] Based on its rumoured success, the mongoose was also introduced to the Hawaiian islands. This aggressive carnivore killed rats but

---

32 Lever 1992, p. 29.
33 Lever 1992, p. 115.
34 Lever 1992, p. 116.
35 Lever 1992, p. 121.
36 Buhôt 1864.
37 *Herpestes javanicus.*
38 Horst et al.2001.

also preyed on pretty well anything else it could catch including lizards, ground-nesting birds and domestic fowl. On the sugar islands of the Caribbean and Hawai'i the mongoose became a pest in its own right. Much later, between 1883 and 1885[39] the mongoose was introduced to Australia to kill rabbits. But Australian rabbit trappers were smarter – they eradicated the competing mongoose in order to protect their means of employment.[40]

In Queensland, instead of rats, cockchafer grubs almost stopped the sugar industry. Cockchafer grubs – cane grubs – are the soil-dwelling larvae of native cane beetles, collectively called scarab beetles. Their larvae normally fed on hardy roots of native plants but, when forests were cleared and cane crops established, larvae could feed on succulent sugar cane roots. They were 'one of the most destructive creatures with which [sugar] planters had to deal'.[41] In the new cane lands, growers had no effective means of controlling beetle larvae or limiting damage to their crops.

In 1891, at Mackay in Central Queensland, damage by cane grubs almost stopped the sugar industry.[42] It was just as bad further south. The sugar company Colonial Sugar Refiners (CSR) hosted a visit by an American entomologist, Albert Koebele, to investigate the problem. This hirsute German émigré travelled to CSR's cane fields south of Brisbane. After inspecting the fields, he concluded that the soil-dwelling larvae, or grubs, of a scarab beetle 'are injurious by eating the roots of the various plants, and … in the absence of any other roots they naturally attack those of the sugar cane'.[43]

Koebele suggested that children collect the grubs but this had already been tried. And CSR also invested in a 'travelling fowl house' carrying about 100 chickens – the caravan was towed behind a horse team to cane fields where chickens were released onto freshly ploughed fields to eat cane grubs. But the chickens could only eat so much, and 'as their appetites became satisfied they would come back to the house and rest',[44] leaving remaining grubs to burrow into fresh soil.

---

39   Funasaki et al. 1988.
40   Peacock & Abbott 2010.
41   Illingworth & Dodd 1921, p. 6.
42   Illingworth & Dodd 1921, p. 7.
43   Koebele 1891.

Toads were another option. Koebele remarked that 'without doubt the presence of toads, if these were introduced, would have a remarkable effect on diminishing the numbers of these [cane beetles] as well as many other injurious insects'.[45] Koebele's liking for toads reflected that of his boss, Dr Charles Valentine Riley, Chief Entomologist for the United States Department of Agriculture (USDA). Riley confessed 'a sneaking kindness for the Toad. He is a sober and quiet philosophical gentleman ... really handsome ... examine his eye; and if you have a jewel[46] about your person that is more brilliant and displays a more tasteful arrangement of colours, you are a fortunate woman'.[47]

Koebele expected toads would catch and eat adult beetles rather than their larvae, cane grubs, that lived in the soil. But CSR did not follow-up the recommendation to import toads. The grub problem continued. The Queensland Government continued to fund the collection of cane grubs and beetles by hand. In the Mackay district in the four years up to 1899, '50 ¼ tons of beetles were collected at an expense of £2,649 19s 7d'.[48] It was good income for small children but ineffective in controlling cane grubs and the damage they caused to crops.

In the closing years of the 19th century, the area of land growing sugar cane was, in Queensland 43,957 hectares (108,535 acres),[49] Hawai'i 51,895 hectares (128,024 acres)[50] and the British West Indies 102,912 hectares (254,103 acres).[51] But this production was put in the shade by Cuba which produced 30% of the world's sugar compared to 7% produced by the entire British West Indies.[52] The old order of world sugar production was changing. Sugar cane grown with irrigation in Hawai'i and grown on newly cleared fertile coastal lands in Queensland was replacing sugar production on exhausted soils in the British West

---

44 Illingworth & Dodd 1921, p. 9.
45 Koebele 1891.
46 Professor Riley may be alluding to Shakespeare's description 'Which like the toad, ugly and venomous, wears yet a precious jewel in its head.' *As You Like It*, act 2, scene 1 (Duke Senior in the Forest of Arden).
47 Walsh & Riley 1869.
48 Illingworth & Dodd 1921, p. 7.
49 Anon. 1933.
50 Dorrance 2001, p. 6.
51 Williams 1970, p. 368.
52 Williams 1970, p. 378.

Indies. In Queensland the sugar industry occupied around a quarter of cultivated land.[53] But the economy of Queensland was severely depressed, the heady growth of the previous decades had stalled. The area under cultivation was stagnant, and the sugar industry was in crisis. A Royal Commission into the industry feared its total extinction because of 'mismanagement, extravagance and inexperience of planters ... financial embarrassment ... losses through unfavourable seasons, disease, exhaustion of the soil, fall in the price of sugar, and loss of confidence in the industry'.[54] The Royal Commission recommended the reconstruction of the industry. Large plantations were broken up to be run by small farmers with access to subsidised central milling. These changes revolutionised the sugar industry and created a class of entrepreneurial cane farmers that characterises the Queensland industry today.

The turn of the century also marked a substantial change in the sugar industry workforce. Since the 1860s, Pacific Islanders, called kanakas – now considered an insulting term – were 'blackbirded' from their homes to work on Queensland's canefields as indentured labour. At the time of Australia's Federation, an alliance of trade unions and anti-slavery campaigners lobbied to repatriate this labour force and to create a White Australia; 'in 1901 a Federal Act provided that no Pacific Islanders should enter Australia after 31 March 1904. Thus was the White Australia policy in regard to the [sugar] industry laid down'.[55]

Captain Louis Hope was among the first to employ Pacific Island labour in Queensland's sugar industry. Through a curious wormhole in time, Captain Hope's nephew, Sir John Adrian Louis Hope, the seventh Earl of Hopetoun and a 'charming but not at all clever'[56] former soldier, became the first Governor General of Australia at Federation. He gave assent to the White Australia legislation that excluded Pacific Islanders. Queensland's flirtation with indentured Pacific Island labour was bracketed by Hope.

The White Australia policy was entrenched further into the sugar industry. In 1902 the federal government provided a financial rebate

---

53   Graves 1993, p. 19.
54   *Report of the Royal Commission into the General Condition of the Sugar Industry 1889*, cited in Johnson 1988, p. 249.
55   Anon. 1930, p. 9.
56   Cunneen 2006.

to growers of white-grown cane. 'White-grown cane means cane in respect of which white labour only has been employed' and 'the expression "white labour" ... is used to the exclusion of all forms of coloured labour, whether aborigines of Australia or not, and whether half-caste or of full blood'.[57] Legislation and financial incentives had the desired effect. 'In 1902, 86 per cent of the sugar produced was produced by coloured labour, but by 1908 white labour was producing 88 per cent of the total sugar produced, and a few years later practically the whole of the sugar was produced by white labour.'[58]

The presence of Pacific Islander labourers – foreign workers – in Queensland was considered a threat to the vulnerable north of Australia and was 'a positive temptation to Asiatic invasion'.[59] Sugar cane was 'the only medium for successfully populating that northern strip of tropical coastline which would otherwise be a source of weakness if not a vulnerable spot in the defence of the country'.[60] By 1930, Queensland cane growers could boast that 'nowhere else in the world is the white man handling tropical production with such success ... It is our vulnerable frontier'[61] and, except for sugar, 'no other industry possessed the same capacity to settle white cultivators on the soil of Australia's vast tropical areas'.[62] If sugar cane was weakened by pests and disease, it weakened northern defences.

But in Queensland at the end of the 19th century it was not foreign labour but rather indigenous pests that were threatening future production and the viability of the industry.

Cane growers, governments and scientists alike, desperate for any form of effective pest control, would embrace the new science of 'biological control'. An era of 'economic entomology' began. It was their time in the sun for entomologists, together with their favourite parasitic and predatory wasps, flies, beetles, moths – and toads.

---

57   Commonwealth of Australia 1902.
58   Anon. 1930, p. 9.
59   *Royal Commission into the Sugar Industry* (1912), cited in Anon. 1936, p. 9.
60   Goldfinch 1935, p. 23.
61   Donald Mackinnon, cited in Anon. 1930.
62   Anon. 1936, p. 7.

5
Ladybird fantasy

Keeping toads in a vegetable garden to consume insects, slugs and snails is the amphibian version of keeping cats to catch mice, dogs to catch foxes, and ferrets to catch rabbits – practices that go back centuries. And it applies to ladybirds. Ladybirds are voracious predators of aphids, scale insects and other pests of crops. They are friends of farmers, orchardists and gardeners. Ladybirds are 'beneficial' organisms that control the activities of other 'pest' organisms. This is called biological control. In the 19th century, in the absence of effective pest control measures, biological control was elevated to the status of a science. At the forefront was the United States Department of Agriculture (USDA), and perhaps the most active proponents in the field were Albert Koebele and his boss, Charles Valentine Riley.

To some, Riley was the 'father of biological control', to others 'Professor Riley',[1] and to those less fond of the restless, ambitious Englishman, 'the General'.[2] With Riley's leadership, beetles, flies, wasps and toads were released on the world; some worked as hoped, some died, and others became pests in their own right. Riley's most notable exploit was using biological control to defeat the cottony cushion scale insect *Icerya purchasi* that was destroying California's orange orchards at the

---

1 United States Department of Agriculture (n.d.).
2 Caltagirone & Doutt 1989, p. 2.

Figure 5.1 Charles Valentine Riley, Chief Entomologist, US Department of Agriculture. (Special Collections, National Agricultural Library. Used with permission.)

end of the 19th century. He recruited predatory ladybirds to control the scale insect and the result was the 'ladybird fantasy'.

Cottony cushion scale insects were introduced accidentally to California on a sprig of Australian wattle in 1868. They were noticed at Menlo Park in 1872, arrived in Los Angeles County by 1876 and had effectively colonised southern California by 1883.[3] Cottony cushion scale insects, each insect about five millimetres long, form colonies that attach themselves to branches and suck the sap; they especially like orange trees. The adults secrete a fluffy white wax, hence the name cottony cushion scale. The fluffy wax covers egg sacs from which nymphs emerge to feed on leaves and stalks. The insects excrete sweet honeydew, leading to infections of sooty mould and infestations of ants; the fruit is ruined.

To solve the problem of the cottony cushion scale, in 1886 Riley placed two USDA officers in California to work under his direction: Albert Koebele and Daniel Coquillett. Coquillett was in charge of gassing insects, and Koebele washed trees with chemicals. Some washes were 'perfectly efficacious and quite within the means of the most indigent [poor] fruit-grower'[4] but growers wanted more effective control and they lobbied the USDA to find a natural predator – a biological control – of the scale insect from its native Australia. The USDA's hands were tied by Congress. In approving the appropriation for Riley to do the work in California, Congress had restricted his efforts to domestic soil. But a small administrative chink became an open gate for Riley. Congress had approved substantial funds for the State Department to exhibit at the Melbourne Centennial Exhibition in 1888. The Chief of the delegation, Californian Senator Frank McCoppin, knew the pest problem first-hand, appreciated the importance of a cure, and simply added Special Agent Albert Koebele to his travelling party and agreed to pay his way.[5]

And so it was that, in mid-October 1888, Special Agent Koebele was seated for afternoon tea in a suburban drawing room in Adelaide, South Australia. This small, busy, 35 year old Prussian-born entomologist considered his surroundings. The already modestly proportioned front room was diminished further by heavily brocaded reddish-brown wallpaper and dominated by a mahogany sideboard hosting stuffed

---

3 Caltagirone & Doutt 1989, p. 3.
4 Willits 1889.
5 Willits 1889.

Figure 5.2 Albert Koebele, HSPA entomologist. (Bishop Museum [SP_207102]. Used with permission.)

and mounted creatures under glass domes. A moth-eaten wedge-tailed eagle centrepiece was flanked by a ringtail possum, a small iridescent kingfisher, a larger kookaburra beginning to moult, and a head of coral.

5 Ladybird fantasy

Gould's prints of Australian birds hung above the sideboard. An immobile and downcast dingo (a native dog) watched over aboriginal artefacts with its one glass eye. A multi-drawered cabinet with inclined glass lid covering pinned butterflies and beetles – an insectarium – stood adjacent. On the wall above, a hand-coloured lithograph of Queen Victoria from her Golden Jubilee reflected in the mirror over the ornate fireplace. On its mantelpiece stood a pair of blue delft vases and a diminishing array of willow-patterned plates. Homely erotica, a barely draped alabaster Aphrodite, waited for Adonis in a corner. And in the lace-curtained bay window, a delicate table hosted regal pelargoniums in a Viennese vase. With pretentions of a drawing room, this was Victorian Adelaide's version of a *kunstkammer*.

Perched on occasional chairs, the master and mistress took tea with their visitors; a damask-clothed table, a triple-storied silver cake stand, a silver tea service, fine gilt-edged bone china.

Lounging expansively on Koebele's left, wearing a fine suit, high collar and ruby-pinned cravat, sat the host,[6] a man of modest means and amateur naturalist. To his left sat the lady of the house, a deal younger than her husband. Her wrists and neck were trimmed with lace where they escaped her day dress, she perched on the front of her chair, her teacup held as a sacred offering. To Koebele's right, in a drab utilitarian suit, sat the elderly Frazer Crawford, photo-lithographer for the South Australian Surveyor General's Department, amateur entomologist and Koebele's contact in Adelaide. Opposite, and with his back to the wedge-tailed eagle fatherly Otto Tepper, natural history collector for the South Australian Museum, fired-up the bowl of his meerschaum, hid the stem in a wilderness of whiskers, and fumigated the room.

Conversation was awkward. Herren Koebele and Tepper in heavily accented English were speaking, Crawford piped a tartan cadence of 'ayes' and 'arrs', Trevelyan spoke to thee and thine in Cornish-English, and only the lady of the house, a second generation colonist, spoke the polyglot of the free settlers of South Australia. This international gathering of adventurers, refugees, economic migrants and travellers

---

6   The meeting was recorded by Koebele but he did not note the name of his host. The host is assumed to be one of the many Cornishmen in Adelaide, recommended by another Cornishman, Mr John Trewenack. Koebele 1890, p. 12.

was common in the British colonies of Australia Felix. They found new benchmarks of success. One such benchmark, unheard of in Cornwall, was the luxury of picking oranges from a tree just outside the back door.

Cottony cushion scale infestations on orange trees was the reason for Koebele's visit. Previously, Crawford had written to Riley offering a fly[7] as a parasite of cottony cushion scale and had sent specimens of the fly to California. Riley's initial brief[8] to Koebele was for him to collect more flies from Crawford and, on 3 October 1888, Koebele had collected the fly from orange trees infested with cottony cushion scale in the garden of another Cornishman in Adelaide, John Trewenack.[9] The scale infestation on Trewenack's orange trees was rare because, in Australia, the scale insect had been well controlled by native parasites. Koebele wanted to find more of the native parasites that were patently so successful. In his search, Koebele attended a meeting of the South Australian Gardeners' Society[10] on 6 October where he delivered a short talk on his work. This stimulated the invitation to afternoon tea.

With his teacup empty, Koebele directed the conversation to the object of his visit – the search for insects – and the party decamped to the garden.

The orange trees were in full sun. Koebele approached a tree with whitish clusters on its branches – it was, indeed, the cottony cushion scale. And for the first time he saw a ladybird eating one of the larger scale insects. Excited, he called the others over and the bowed heads of two Prussians, one Scotsman and a Cornishman, watched in silence as the ladybird – the vedalia beetle *Vedalia cardinalis* (now *Rodolia cardinalis*) – went about its business.

> A few moments were spent in cleaning herself and then ... she attacked and devoured a half-grown scale [insect]. This was eaten into from the back, very quietly at first, yet in a little time she became lively, almost furious, tearing the scale off from its hold by the beak and turning it up and down in the air with the mouth parts, assisting

---

7   The fly was called *Lestophonus* by Koebele (Koebele 1890, p. 11). The fly was reported as *Cryptochaetum iceryae* by Debach (Debach 1974, p. 92).
8   Koebele 1890, p. 7.
9   *The Adelaide Observer*, 13 October 1888.
10 *The Adelaide Observer*, 13 October 1888.

5 Ladybird fantasy

Figure 5.3 Vedalia beetle eating cottony cushion scale. (www.agefotostock.com AGS-153653-S-5736. Used with permission.)

in this with the anterior legs. In about one minute this was devoured and nothing but the empty skin left, after which she went to work, business-like, and deposited eggs quietly, sitting at rest upon the scales, and every few minutes thrusting an egg in between or generally under them.[11]

Koebele collected the scale insects together with both their fly parasites and ladybird predators and readied them for shipment back to California. Koebele contacted a fellow Prussian, the elderly Mortiz Schomburgk, Curator of the Adelaide Botanic Gardens, who had his men construct a Wardian case – a miniature glasshouse – in which he placed live orange trees and *Pittosporum* shrubs in pots. Koebele then introduced the scale insects, parasites and predators to the case and dispatched it to Coquillett, his USDA colleague in Los Angeles. But the Wardian case was 'handled in such a rough manner by the steamer hands at Sydney'

---

11   This description by Koebele was originally made of ladybirds in a glass jar: Koebele 1890, p. 31.

that the case was damaged and ladybird larvae were found wandering around the outside of the case when it arrived. Koebele sent the next two consignments well wrapped and in the ice room of the steamer under the watchful eye of the ship's butcher.[12]

Between November 1888 and January 1889, Coquillett received three consignments totalling 129 ladybirds. He placed them in a tent – an insectary – constructed over an orange tree owned by Mr Wolfskill outside Los Angeles. By mid-April, all the scale insects on the tree had been eaten and the tent was full of hungry ladybirds, so Coquillett simply opened one side of the tent and released them into the orchard.[13] Word got around the neighbourhood about the new ladybirds and farmers took infested branches from their own trees to the orchard, laid them on the ground for two hours, then took them back, covered in ladybirds, to their own orchards.[14]

The ladybirds were an outstanding success. By mid-June 1889, Coquillett had released 10,555 ladybirds to 288 orchardists in the Los Angeles region. From there, the ladybirds 'distributed themselves of their own accord'.[15]

Over several months, Koebele sent 40,000 beetles of 40 species of ladybird from Australia. But only four species of ladybird became established in California, and just one species, the vedalia beetle, successfully controlled the cottony cushion scale.[16] By contrast, the parasitic flies fared poorly – the ladybirds seemed to have eaten so many scale insects that there was very little left for the flies to parasitise.

Flies and ladybirds alike were released into the wild in a huge uncontrolled experiment. There was no prior testing to see what else they would feed on. No testing to see if they would become pests in their own right. Their progress was observed, but only if they didn't travel too far, and no detailed records were kept of their progress.

It could have gone so dreadfully wrong, but it didn't.

Orange orchards recovered. At the peak of scale infestations, only 781 boxcar-loads of oranges were railed from Riverside, but in 1891,

---

12 Koebele 1890, p. 14.
13 Coquillett 1889.
14 Caltagirone & Doutt 1989, p. 5.
15 Coquillett 1889.
16 Caltagirone & Doutt 1989, p. 10.

5 Ladybird fantasy

two years after the introduction of ladybirds, almost three times as many oranges were consigned.[17] Koebele was feted by citrus growers and 'given a gold watch and diamond earrings for his wife'.[18] It started a world-wide craze for ladybirds – the 'ladybird fantasy'.[19]

From California, vedalia beetles were distributed far and wide: to Egypt, Portugal, Cyprus, the Soviet Union, Puerto Rico, Venezuela, Uruguay, Peru, Chile, Argentina, Hawai'i, the Philippines, Guam, and Palau.[20] And the ancestors of virtually all these colonies of vedalia beetles was the 129 adults collected by Koebele in Adelaide, South Australia. In their new homes they were released into the wild in the same manner that Coquillett had released them in California. If it was good enough for the USDA, it was good enough for everyone else.

The ladybird fantasy was understandable. In the late 1880s, farmers had no effective agricultural chemicals to control insect pests.[21] Biological control was a necessity. Vedalia beetles heralded 'the start of modern classical biological control – the importation and release of an organism outside its natural range for the purpose of controlling a pest species'.[22] Entomologists, marking the centenary of biological control in 1989, declared that 'no single achievement has more thoroughly, soundly, and significantly established a major pest control tactic than the vedalia project. All subsequent projects, programs, advances, and refinements in the theory and practice of biological control have sprung from this single event'.[23]

Imitations of the vedalia program had far greater ecological consequences than any entomologist could have comprehended at the time. Under the USDA's banner, exotic beetles, wasps, flies, bugs and toads

---

17 Lelong 1900, p. 20.
18 Caltagirone & Doutt 1989, p. 8.
19 Caltagirone & Doutt 1989, p. 10.
20 Caltagirone & Doutt 1989, p. 13.
21 It was not until the late 1930s that broad-spectrum organochlorine insecticides, most notably DDT, aldrin and dieldrin, and the more acutely toxic organophosphate insecticides like parathion and malathion, came on the market. But they left their own persistent and lethal legacy on the environment before they were banned.
22 Howarth 1991.
23 Caltagirone & Doutt 1989, p. 14.

were released unregulated and mostly unrecorded into new environments. It turned into a huge gamble with the environment.

It was not science.

While working with the USDA, Koebele introduced many new insects to Hawai'i. It became something of a love affair with the islands. He was very busy and, although most of his introductions went unrecorded, there are some records. In 1889 he introduced the ichneumonid wasp *Pseudamblyteles koebelei* from California to control the moth *Spodoptera exempta* whose larvae were defoliating pasture grasses.[24] In 1890, he introduced the vedalia beetle *Rodolia cardinalis* from California to control cottony cushion scale.[25] In 1891–92 it was more ladybirds from Australia to control the pink sugar cane mealy bug *Saccharicoccus sacchari* and other scale insects.[26] And in 1894 it was another Australian ladybird *Coleophora inaequalis* introduced to control the aphid *Longuiungis sacchari* that sucked the sap of sugar cane.[27]

The love affair matured into a marriage. In 1893 Koebele resigned his position with the USDA to become State Entomologist with the Hawaiian Board of Agriculture and Forestry where he was free to carry on the wholesale importation and release of exotic insects.

Unfortunately for Charles Valentine Riley, Koebele's former boss, a close encounter with geology in 1895 cut short his life. When cycling down a hill at great speed, he parted company with his bicycle when it struck a granite block. He fractured his skull and never regained consciousness. He left behind a wife and six children, one named Cathryn Vedalia Riley.

Before his death, Riley proposed the model of Experiment Stations funded for research into the 'efficient production, marketing, distribution, and utilization of products of the farm as essential to the health and welfare of people and to promote a sound prosperous agriculture and rural life'.[28] Riley's USDA Experiment Stations helped settlers develop new lands for farming in America. In Hawai'i, sugar planters funded developments along the same lines. In 1895, the eight dominant

---

24 Pemberton 1964, p. 698.
25 Pemberton 1964, p. 699.
26 Pemberton 1964, p. 702.
27 Pemberton 1964, p. 704.
28 United States Department of Agriculture (n.d.).

5 Ladybird fantasy

sugar planters on the islands changed the structure and the name of their organisation – The Planters' Labour and Supply Company – to The Hawaiian Sugar Planters' Association (HSPA). The HSPA then established its own Experiment Station.

In Hawai'i, as in Barbados, continuous cropping with sugar cane had depleted soil fertility and structure. Because of this, HSPA determined that the work of the Experiment Station was to be in agricultural chemistry, advising planters about nutrition of their crops, and they wanted an agricultural chemist to lead it. Through links with Louisiana sugar planters, HSPA contracted the services of Dr Walter Maxwell as Director of the Experiment Station. He arrived in Honolulu in April 1895. Maxwell had an impressive pedigree: five years in Germany working on soils for sugar beet, four years in Washington and the USDA sugar station in Nebraska, and two years in Louisiana as professor of chemistry working with sugar.[29]

In 1898 the USA had a 'splendid little war' with Spain lasting just four months.[30] It gave America new colonies of Cuba and Puerto Rico in the Caribbean, and the Philippines and Guam in the Pacific. Despite appearances, President McKinley declared that the American flag had not been planted in foreign soil to acquire more territory, but for humanity's sake. Nevertheless, the USA also annexed Hawai'i in the same year. Sugar planters were handed a wealth of new territories – and new pests to go with them. The USDA opened the Puerto Rico Agricultural Experiment Station at Mayagüez in 1901, and the Hawaiian Agricultural Experiment Station in 1902.

On Hawai'i, the USDA had to deal with island landscapes denuded by wholesale clearing and export of sandalwood 80 years earlier, and further destruction of vegetation by unregulated grazing by cattle and sheep. According to the British entomologist David Sharp, 'the land fauna was known to be undergoing great impoverishment'.[31] Sharp coordinated the *Fauna Hawaiiensis or the Zoology of the Sandwich (Hawaiian) Isles*[32] for the Royal Society of London and the British Association for the Advancement of Science. The collection of more

---

29 Grammer 1947.
30 Bethell (n.d.).
31 Sharp 1913.
32 Sharp 1913.

than 100,000 animal specimens was an heroic undertaking, carried out between 1892 and 1903 by a young English zoologist and Oxford graduate, Robert Perkins.

On field work in Hawai'i, Robert Perkins often travelled alone for many months at a time, carrying his equipment, clothing, a small tent, an oil stove, and a simple diet of rice, 'coffee and sugar and one or two kinds of tinned meats ... sometimes for weeks together I never saw a human being. Occasionally a native would come up and leave letters for me or take them down from my tent'.[33] In 1896, Koebele accompanied Perkins on a collecting trip. Perkins recalled,

> After spending some weeks in the high, wet forest of the windward side [of Maui], we left our tent, and carrying as few *impedimenta* as possible, we worked about the summit ... sleeping in the open or in such natural shelters as were available. Being lightly clothed we were a good deal troubled by the sharp frosts of a night ... [this was] the first of several hard trips we made together.'[34]

But, 'for the most part he [Koebele] was absent on economic[35] [entomology] work in other countries'.[36]

Koebele's unrecorded importations worried Sharp who had seen first-hand the collapse of simple natural ecosystems on Hawai'i and was concerned about the impact of imported insects on what was left of island flora and fauna. Perkins had also seen the impact of Koebele's importations and discussed with Koebele 'the possibility of his economic introductions proving antagonistic to my own work on the native fauna, especially as some of the rarer species of native hemerobiids [brown lacewings] ... had disappeared after the introduced *Coleophora* [moth] had eaten up the *Aphis* [that formed the lacewings' diet]'.[37] Sharp complained to Leland Howard, Riley's successor, about

---

33 Sharp 1913, p. xxxviii.
34 Perkins 1943.
35 Economic entomology was to do with management of insects as distinct from taxonomic entomology which was to do with collecting, describing and classifying insects.
36 Sharp 1913, p. xvi
37 Perkins (n.d.).

5 Ladybird fantasy

Figure 5.4 Robert Perkins, HSPA entomologist. Photographed at Dartmoor, England. (Bishop Museum [SP_53672]. Used with permission.)

Koebele's unregulated and unrecorded activities: 'It is important that a permanent record shall be secured of what Mr Koebele has done ... Mr Koebele is actually making a huge biological experiment.'[38]

Howard was blunt with Sharp. Howard's philosophy was that 'Man is the dominant type of this terrestrial body, he has overcome most opposing animal forces, he has subdued or turned to his own use nearly all kinds of living creatures.'[39] For Howard, Hawai'i was purposefully the centre of a 'huge biological experiment'. Koebele continued, unconstrained. In 1902 he introduced 23 species of insects from Mexico to control the invasive *Lantana* weed. This was an innovation: 'the first time the control of noxious plants with insect enemies was ever attempted anywhere in the world'.[40] Again, there was no prior testing and no recorded observations.

But Koebele's work was soon supported by a team of entomologists employed in finding biological control agents for pests of sugar cane. A new pest had arrived in Hawai'i; it was discovered by Perkins. Towards the end of 1900, Perkins was working late one night in Waialua, Oahu, window open, light on, a sea breeze fluttering the curtain. Among the insects attracted by the light he noticed something new, a species of leaf hopper that he had not seen before. He researched his new find and discovered that it had been introduced to Hawai'i on cuttings of cane imported from Queensland, Australia.[41] A year later, the leaf hopper was causing serious damage to Oahu's cane crops and by the middle of 1903 it was found on all the Hawaiian islands. Sugar yields were falling and the leaf hopper threatened the future of the industry. The leaf hopper was later named after its finder – *Perkinsiella saccharicida*.

Perkins ceased collecting for the *Fauna Hawaiiensis*[42] and crossed-over to the dark side. In 1904 he was appointed Director of the new Division of Entomology of the HSPA Experiment Station.[43] Perkins became an economic entomologist in the tradition of Koebele, importing and releasing insects for biological control. It was an anathema

---

38 Howarth 1991, p. 486.
39 *The New York Times*, 28 December 1921, p. 6.
40 Pemberton 1964, p. 719.
41 Pemberton 1948, p. 34.
42 Sharp 1913, p. xxxv.
43 Perkins & Kirkaldy 1907.

to his former mentor, Sharp, the coordinating entomologist for the *Fauna Hawaiiensis*. The target for the new Division of Entomology in the HSPA was the sap-sucking sugar cane leaf hopper: 'damage was so extensive that whole [cane] fields of great area were practically killed outright'.[44] Perkins was assisted by five entomologists including his old colleague, Albert Koebele. The following year, English-born Frederick Muir joined the team. Muir was a son-in-law of Sharp[45] but did not share his father-in-law's reservations about biological control.

In 1904, Perkins and Koebele travelled to Queensland to search for predators or parasites of the leaf hopper. There they found the eggs of the leaf hopper parasitised by a tiny mymarid wasp *Paranagrus optabilis*. The wasps, each less than half a millimetre long, were borne aloft on lacy wings. They were sent back to Hawai'i for release into cane fields to parasitise eggs of the leaf hopper. The following year, Koebele and Muir travelled to Fiji and sent more parasitic wasps back to Hawai'i.[46] Koebele then went on vacation but still sent parasites of leaf hoppers back to Hawai'i from Arizona and from Germany.

Koebele's practices showed scant regard for consequences. His method was to collect first and ask questions later. In 1890, reporting on his collecting trips for the vedalia beetle, Koebele wrote

> I had no time while in the field to study much of the life-history of this valuable insect. My first motto was always 'get as many as possible.' If once established here [in California], the life history may be studied at leisure.[47]

And he left a scant trail of evidence. Perkins wrote of Koebele,

> he was not a great reader of entomological literature ... he published no notes of a systematic nature ... excepting some official reports and even these were to him an uncongenial task.[48]

---

44   Pemberton 1964, pp. 689–729.
45   Crook 1932, p. 114.
46   Pemberton 1964, p. 708.
47   Koebele 1890, p. 30.
48   Perkins (n.d.).

Figure 5.5 Frederick Muir, HSPA entomologist. (Bishop Museum [SP_53672]. Used with permission.)

5 Ladybird fantasy

The introduction to Hawai'i of parasitic insects from America, Europe, Asia and Australia was spectacular in both scope and scale. It was a 'huge biological experiment' under the guise of biological control. Industry members of the HSPA got a spectacular return on their investment but there was no record of the collateral damage to the fauna of Hawai'i. The increase in the number of economically beneficial insects was an end in itself.

In 1904, soon after joining the HSPA Experiment Station, Frederick Muir took the reins of another biological control program. He was to find new parasites of an old pest of sugar cane in Hawai'i, the sugar cane weevil borer *Rhabdoscelus obscurus*, introduced to Hawai'i from Tahiti back in 1865.[49] With the encouragement of his employer, this professional scientist was absent from his desk for four years travelling the world collecting insects. Muir's exploits became one of the enduring legends of biological control in sugar cane.

Muir's epic began in 1906 with a hunt for the beetle in China. Muir joined his English compatriot, entomologist John Kershaw, in Macau, in a colonial villa on the tree-lined crescent of the Avenida da Praia Grande, overlooking the bay.[50] Kershaw had just published his *Butterflies of Hong Kong* and was delighted to discuss entomology with a fellow Englishman. The pair set off from Macau equipped in the manner of the modern entomologist:

> a cotton umbrella, and strong and narrow steel trowel or digger, a haversack slung across the shoulders, a cigar box lined with sheet cork, and a small knapsack attached to a waistbelt which girts a coat, not of many colors, but of many pockets, so made that in stooping nothing falls out of them. The umbrella is one of the indispensables. It shields, where necessary, from old Sol's scorching rays and from the pelting, drenching storms; brings within reach, by its hooked handle, many a larva-freighted bough … and forms an excellent receptacle for all insects that may be dislodged from bush or branch.[51]

---

49 Pemberton 1964, p. 697.
50 Easton 2008, p. 2,085.
51 Riley 1892, p. 26.

And there was

> The problem of the influence of the sun's rays upon the body ... A good light sola pith helmet with special protection for the back of the neck is the best thing that can be employed ... light in weight and colour, red or orange lining being also an advantage ... in the protection of the body from the actinic rays of the sun.[52]

This was worn together with a spinal pad 'to protect the back of the neck and the upper part of the spine [from the sun]' and a cholera belt, 'a broad flannel belt, fixed by two buckles ... worn outside the vest or shirt ... to protect the vital organs of the abdomen, from a tendency to chill, which may easily occur in the Tropics in sudden changes of temperature'.[53]

Thus equipped, Muir and Kershaw travelled up the Pearl River to Guangzhou and on to How Lik Monastery near Guangdong. For six months, the pair inspected crops and vegetation, masquerading as doctors searching for medicines. They did not find the beetle, but they sent many new insects back to Hawai'i for release into cane fields in the hope that some of them might be useful. For most of the following year the pair searched the Malay Peninsula and the islands of Borneo, Java and the Moluccas, and most of Wallace's Malay Archipelago[54] until, in a sago-palm in Amboina, they found a beetle similar to the cane borer. The beetle hosted a parasitic fly. They were two years into their journey.

In Amboina, Muir, despite being almost blinded by a spitting cobra, collected as many beetles and parasites as he could, but the insects died on the long boat trip to Hong Kong where Kershaw had planned to establish a breeding colony. Undaunted, in 1908 Muir returned to where he first found the beetles but, searching high and low in the sago palm swamp, he could find no more. He then turned to the island of New Guinea where he found a beetle hosting the parasitic tachinid fly *Lixophaga sphenophori*. Tachinid flies are similar to house-flies except that they are more bristly. They lay their eggs on or near other insects,

---

52   Harford 1911, p. 13–14.
53   Harford 1911, p. 13–14.
54   Wallace 2008(1890).

and their larvae burrow into their hosts and devour them from the inside.

Muir decided to breed these flies on their hosts, the beetles – this time in Brisbane, Queensland. But he 'fell ill with typhoid fever and lay flat on his back in an Australian hospital for five weeks. While he was in hospital, disaster struck his collection: his parasites hatched out and died'.[55] Muir returned to New Guinea in February 1910 where, weak with fever, he found and collected more parasites. He fell ill again and called on the insect rearing skills of his friend, Kershaw. At the sugar mill in Mossman, north Queensland, Kershaw set up a breeding facility for Muir's beetles and flies. The flies bred successfully. Kershaw had skilfully constructed the breeding cages more than two metres high – high enough to accommodate the flies' pre-nuptial flight essential for successful mating. Then Muir established an intermediate breeding colony of beetles and flies in Fiji in order to get them back safely to Hawai'i.

In late 1910, almost four years after setting out, Muir and Kershaw reached the HSPA Experiment Station in Honolulu with live tachinid flies to parasitise the Hawaiian cane borer. Tachinid flies were successful parasites, adopted by sugar growers around the world, and added to the legendary record of the HSPA in biological control.

Remarkably, Muir was absent from his workplace in Honolulu for four years of travel, salaried and funded while searching for beneficial insects. There could be no stronger evidence than this of HSPA's corporate zeal to pursue agents of biological control.

Ten years later, on yet another trip to Queensland, Muir discovered the mirid bug *Tytthus mundulus* sucking the life juices out of leaf hoppers. This mirid bug had the greatest success of all the insects introduced to Hawaiian cane fields and the leaf hopper control program was hailed another seminal case of biological control in the sugar industry.[56]

But Sharp was right – Koebele, followed by Perkins and Muir, conducted a huge biological experiment on Hawai'i. And this experiment continued, apparently, without recorded observations. It may have been a huge biological experiment, and it wasn't science, but cat-

---

55 Easton 2007.
56 Pemberton 1964, p. 708.

tlemen, horticulturalists and cane growers were most grateful for the HSPA's efforts.

Dr Walter Maxwell lasted only five years as Director of the HSPA Experiment Station in Hawai'i. In December 1900, he was lured away to Queensland, Australia, to start another new Experiment Station, the Queensland State Government's Bureau of Sugar Experiment Stations (BSES). His appointment came about because James Chataway, Mackay sugar planter and Queensland correspondent for the *The Louisiana Planter and Sugar Manufacturer*, learnt of Maxwell's reputation. Later, as Minister for Agriculture and Stock in Queensland, Chataway brought Maxwell to Queensland to evaluate the state of the sugar industry.[57] Chataway appointed him foundation Director of BSES. HSPA members gave Maxwell a backhanded congratulation, remarking 'The sugar industry of that colony [Queensland] has for several years been in a very demoralised condition, and it will need most heroic measures on the part of the Colonial Government to redeem [it] … if any man can do it, Dr Maxwell can.'[58]

Albert Koebele retired to his homeland in Germany. His legacy, the practice of biological control as demonstrated in Hawai'i by his disciples Perkins and Muir, became a benchmark for the world. Koebele's ladybirds were eventually distributed to more than 50 countries[59] and, following in this tradition, toads would soon join the ranks of wasps, flies, bugs and mongoose as agents of biological control.

Soon, millions of young men, including scientists from the USDA, HSPA and BSES, would embark for Europe as nations were sucked into the vortex of the Great War – it heralded the end for Koeble. This German émigré, creator of the ladybird fantasy, hero of Californian citrus growers and of Hawaiian cane growers, was denied an American visa. His health declined and he never again contributed to the huge experiment of biological control.

---

57   Anon. 1900b.
58   Anon. 1900a.
59   Sweetman 1958, p. 3.

6
The cane beetles

Autumn 1918, Western Europe. At Waldkirch in Germany's Black Forest, Albert Koebele, entomologist, champion of biological control, sat racked by fevers, malarial parasites in his bloodstream. He contemplated his own mortality and his country's imminent defeat. Not far to the west, soldiers of the American Expeditionary Force from his adopted country joined French, British and Commonwealth troops on the Western Front. Young soldiers from all walks of life volunteered, among them cane farmers and scientists from the sugar industry. George Wolcott, 29 year old entomologist with Koebele's own former employer, the United States Department of Agriculture (USDA), was from upstate New York via the cane fields of Puerto Rico. His artillery company faced off against Koebele's countrymen outside Verdun.

Australians were well represented. It was a most loyal British dominion. In 1914, earnest young Queenslanders from the cane fields of Gordonvale enlisted for 'the stunt' under the banner of 'The Cane Beetles'.[1] For Alan Dodd, assistant entomologist in Queensland's Bureau of Sugar Experiment Stations (BSES), cane beetles were his business. When war broke out, 20 year old Alan followed his two elder brothers and enlisted as a non-combatant; it 'seemed the only thing to do'.[2] He served as a medical orderly in the 15th Field Ambulance on the

---

1 Morton 1995, p. 82.
2 Dodd 1945(1917)a.

River Somme.³ His field ambulance immediately behind the front line patched men and got them to better treatment, but one in four went to be buried. Unending ranks of young men wearied and grown old. A cricketer's arm, a stockman's legs, feet a mother bathed at birth, a soul stripped of remorse. All left in the soil of France. Bodies, boots and brasses embraced by clay. But solace for this tall, gangly entomologist was the soft green cane fields of Gordonvale, the landscape of his mind's eye, where he walked in his dreams.

In 1918 the insect that most occupied this entomologist was the human body louse *Pediculus humanus*. Like cane beetles, the largest of them were called greybacks; 'a fully grown one being about three sixteenths of an inch [five millimetres] long and in the grey army shirts and singlets they almost defy detection ... they are a thousand times worse than rabbits at breeding'.⁴ There was debate in the trenches about how best to kill them; you could either kill the old ones and let the young ones die of grief, or kill the young ones and catch the old ones when they went to the funeral.⁵

Nearby, Arthur Bell, junior laboratory analyst with the Queensland Department of Agriculture and Stock, served as a gunner with the 12th Army Brigade Australian Field Artillery.⁶ Home for this brown haired, blue eyed 18 year old was the patchwork-green market-gardens of Laidley, west of Brisbane – a land of plenty compared to the Western Front, which was 'the most frightful picture of desolation. Nowhere was there a living soul to be seen. Great bare hills ploughed into a wilderness of shell-holes, a fine, grey misty rain, not a house – not even a tree'.⁷ Bell arrived just two weeks before the Armistice on the sort of day Eric Remarque committed to literature: 'so quiet and still on the whole front, that the army report confined itself to the single sentence: all quiet on the Western Front'.⁸

---

3   Dodd 1916.
4   George Rayment describing lice in the trenches at Gallipolli. Cited in Hamilton 2004, p. 210.
5   Graves 2009(1929), p. 89.
6   Bell 1917.
7   Glubb 1978, p. 204.
8   Remarque 1993(1929).

6 The cane beetles

By the spring of 1919 field guns had been limbered-up, gassed and maimed despatched to sanatoriums, field hospitals packed away, cemeteries gridded out. Soldiers' lasting memories of Flanders that spring were red poppies carpeting the blasted wasteland. A generation wasted. No 'passing-bells for those who die as cattle', but 'their flowers the tenderness of silent minds, and each slow dusk a drawing-down of blinds'.[9] Their loss felt as keenly in vineyards on the Rhine as in corn fields in Kansas, hop fields in Kent and cane fields in Queensland.

Those who lived returned home with unseen wounds. George Wolcott would return to the insects of the Caribbean, like a moth, never settling for long. Soon after returning, he would promote the cane toad as the saviour of the sugar industry. He and Arthur Bell would cross paths in Puerto Rico and Bell would be the one to propose that the cane toad be brought to Australia. Alan Dodd, morose and solitary, would never marry. He would continue his love of insects and lead one of the world's most acclaimed applications of biological control, the attack on prickly-pear in Australia.

The experiences of these young men on the Western Front were to mark them for ever. Returned soldiers imagined corpses lying on the street, strangers would assume the faces of dead comrades, landscapes of home would be scanned for defensive positions, and shells bursts destroyed sleep.[10] The war's impact on their generation would be profound.

At war's end 167,000 Australians were scattered across France, Belgium, England and Egypt, eager to return home. Eighteen months were needed to repatriate them all, so the Australian Government set up a 'scheme of "non-military employment" to include training for almost every form of civil occupation'.[11] Under this scheme, Arthur Bell studied chemistry in Manchester and Alan Dodd was employed in the entomology section of the Natural History Museum in London. After risking their lives to defend the British Empire, it was a gift for these two young men to work in their chosen professions, away from the front lines.

---

9   Owen 1966, p. 81.
10   Graves 2009(1929).
11   Scott 1941, p. 827.

Figure 6.1 Alan Dodd. (Dodd family. Used with permission.)

Alan Dodd, his three brothers and two sisters were reared among insects. His father, Frederick, a former bank clerk, was the 'Butterfly

Man of Kuranda'[12] in north Queensland, collecting tropical butterflies, moths and beetles and taking his beautiful displays on national tours. Alan, Frederick's third son, went to work with insects in BSES at Meringa just down the Atherton Range from Kuranda. Before enlisting, Alan worked under the mercurial entomologist Alexandre Girault from the University of Illinois. On his return from the Western Front, Dodd had a new mentor, James Illingworth, a Stanford and Cornell trained entomologist recruited from the College of Hawai'i.[13] In 1919, Illingworth gave Dodd a job to suit this returned soldier. From evening, through the night and into early morning, for two successive summer growing seasons, Dodd, mostly alone, sometimes with Illingworth, observed the movements and mating habits of cane beetles. He was stationed on CSR's cane farm at Greenhills near Gordonvale. It was one of the most badly grub-infested cane farms in the district. Nothing like this had been done before. Dodd's records became vital in understanding the behaviour of cane grubs and adult beetles.

January 1920, North Queensland. On a humid midsummer evening a lone entomologist waited in the cane fields. It had rained late in the afternoon and the still air among the two metre high cane stalks carried the heavy scent of moist soil and fungal spores. After sunset, a cloudless night and a monochrome landscape under moonlight. Dodd scanned the terrain. To the south, the forested slopes and pyramidal peaks of Bellenden Ker were capped by thin cloud. To the west he could make out the dark outline of the Atherton Tableland. To the east lay steep, forested coastal hills. Away to the north cane farms quilted the valley.

Dodd did not appreciate moonlight. His experience was that 'instead of romance it brings bombs.'[14] In spite of the moonlight it was a defensible position with Lewis guns on the perimeter, 'The Cane Beetles', bayonets fixed, on the parapet awaiting the whistle, and Dodd, medical orderly, waiting for the bombardment. Waiting for the inevitable call for stretcher bearers and 'the carry', struggling through mud with 'a shattered mass of flesh and blood.'[15] But just now the still air

---

12 The Butterfly Man of Kuranda (n.d.).
13 Griggs 2005, p. 6.
14 Dodd 1945(1918).
15 Dodd 1945(1917)b.

carried no sound until a stone curlew cried ker-*looo*, ker-*looo*, distressingly human, a cry from no-man's-land, from beyond the wire. Then the racket started: 'Half an hour after sunset swarms of [French's cane] beetles rise out of the ground simultaneously on every side … as far as the eye can see … they constantly strike against the [cane] leaves … very noticeable above the constant hum of the seething swarm'.[16] The wild night flight of cane beetles as they emerged from the soil lasted for only 10 minutes, after which they began to settle on leaves, wire fences, anything. Dodd squatted to avoid the swarm and fumbled in his shirt pocket for a match. Still wary of a sniper's bullet, he struck the match, lit the small oil lamp, shielded the light and watched the beetles copulate. 'The females always settle first, but each is quickly surrounded by several males … the attached male always lets himself fall back as soon as union is secured, and hangs head downward supported only by the genitalia'.[17] After these sexual gymnastics, the male pulls himself free and flies away to bond with other males in the feeding trees, abandoning the impregnated female alone in the cane.

The greyback cane beetle is larger than French's cane beetle and just as committed to copulation as soon and as often as possible after leaving the soil. For the greyback cane beetle,

> copulation takes place repeatedly as long as the beetles are on the wing, and they are polygamous … beetles mate repeatedly, evening after evening, beginning as soon as dusk comes on, during the whole of their aerial existence, the males going from one female to another … [and] they continue copulating even when the females are packed full of ripe eggs which are ready to lay.[18]

After spending the night clustered on tree trunks, females return to moist soil to lay their eggs. The pre-dawn flight of females returning to the cane fields is just as remarkable as the evening swarm. Dodd in position, ready for action, watched as they stormed the cane fields.

---

16   Illingworth & Dodd 1921, p. 45.
17   Illingworth & Dodd 1921, p. 45.
18   Illingworth & Dodd 1921, p. 15.

6 The cane beetles

> At exactly 5 o'clock [in the morning] I saw the first one coming from where the sun was soon to appear, a little to the south of east. In a few minutes dozens were coming, all from the same quarter ... and they went as far as the eye could see against the skyline to the west and north-west ... The morning flight constantly ended before 5.30, and the air became noticeably quiet.[19]

It was the most thorough research on cane beetles and grubs since Captain Louis Hope planted the first crop of cane at Ormiston 60 years earlier. The research monograph written by Illingworth and Dodd made the first 'definitive observations of the mating habits of the greybacks'.[20] For anyone reading the monograph, it is clear that biological control of cane beetles would require a very special predator. Each day of the summer months, but only after rain, predators would have to be ready for two short half-hour periods, one just after dusk and the other just before dawn. There were no other times to attack. The beetles would have to be caught in the few seconds that elapsed either at dusk as the beetles emerged from the soil and took wing, or before dawn when the beetles landed and in the few seconds before they entered the soil. And predators would have to consume beetles as if there was no tomorrow – for what would they eat tomorrow once the beetles were gone?

Good science is often pedestrian. The methodical and unglamorous observations by Illingworth and Dodd showed the impossible task confronting a predator species if it was to control populations of the many species of cane beetles and their grubs in the cane fields. But they offered no solution, concluding instead that 'it would appear that man, by his serious interference with the balance of nature, is mainly responsible for much of the devastation that is now resulting to his crop [of sugar cane]'.[21]

It was not a popular conclusion among cane growers.

Entomologists have now identified about 120 species of scarab beetles in Australia. The larvae (grubs) of 19 of them are pests of sugar cane. Of these, the greyback cane grub *Dermolepida albohirtum*, and

---

19   Illingworth & Dodd 1921, p. 16.
20   Illingworth & Dodd 1921, p. 15.
21   Illingworth & Dodd 1921, p. 14.

French's cane grub *Lepidiota frenchi* are the most damaging pests in north Queensland. The Childers cane grub *Antitrogus parvulus*, negatoria cane grub *Lepidiota negatoria* and the southern one-year cane grub *Antitrogus consanguinensis* create most damage to cane in southern Queensland. And the rhopaea cane grub *Rhopaea magnicornis* is the greatest pest in New South Wales cane fields further south.[22] Life cycles also vary from north to south. In the north, greyback cane grubs have a one-year life cycle. In the south Childers cane grubs have a two-year life cycle and other cane grubs living between these two have either one or two-year life cycles. Thus, in summer in a cane field, there may be several different species of beetles in different developmental stages, from pupae deep in the soil, to cane grubs eating roots, and to adults on the wing. It is a complex problem and entomologists still consider that greyback cane grubs and their allies 'are the most important group of pests in the Australian sugar industry'.[23]

Today, cane grubs and beetles are addressed through systems of integrated pest management that take into account pests, predators, parasites, seasons, soils, crop stage and available pesticides. But in 1921, the best that BSES could recommend was to apply arsenic to the soil for the grubs to ingest, or to dust the beetles' feeding trees, often gully rainforest remnants, with either calcium arsenate or lead arsenate.[24] Even poisoning the soil or the surrounding tropical environment with arsenic would not eliminate the grub problem and there was little understanding of the impact of persistent poisons on the tropical environment. In 1921, biological control – importing a foreign predator or parasite – was the ultimate technology in crop protection, and the last resort for cane growers.

Illingworth and Dodd knew that Koebele had recommended toads to control scarab beetles when he visited CSR's cane fields in 1889. In their monograph they repeated Koebele's recommendation that 'without doubt the presence of toads ... would have a remarkable effect on diminishing the numbers of these [cane beetles]'.[25] Koebele's reputation ensured that his advice was acknowledged but, curiously, in this

22   Allsopp 2001.
23   Allsopp 2001.
24   Illingworth & Dodd 1921, p. 94.
25   Koebele 1891.

6 The cane beetles

  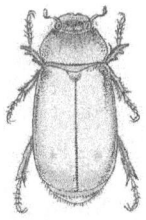

Figure 6.2 Larva, pupa and adult greyback cane beetle. (Redrawn from Illingworth & Dodd 1921. Australian sugar cane beetles and their allies. Division of Entomology, Bulletin 16. Bureau of Sugar Experiment Stations, Brisbane.)

instance his advice was not heeded. Perhaps the futility of the toad's task was obvious or perhaps, post-war, the opinion of the German was discounted. That was certainly the case in Hawai'i in 1918 where the opinion among cane growers, expressed in the *Hawaiian Planters' Record* and read widely in Australia, was that 'We used to let Fritz [Germany] do a great part of our scientific research for us ... our inertia made Fritz believe that he was the only one that could do this thing and therefore that he could lick all creation ... it is better for everybody if we do our own research, and that we get to work at it right away.'[26]

In 1921, research into cane beetles was cut short. Both Illingworth and Dodd left BSES. Illingworth returned to Hawai'i to join the Bishop Museum and Dodd was recruited by the Commonwealth Prickly-pear Board to help solve Queensland's most significant agricultural problem, the unchecked spread of prickly-pear cactus. It was a problem of far greater magnitude and economic consequence for the State of Queensland than cane grubs, and there were too few entomologists to go around. Dodd would join a research program to eradicate prickly-pear that had been running for nine years. Under his leadership the prickly-pear eradication program would become a seminal example for biological control in Australia, and later the world. The program followed strict protocols that minimised the risk of any imported biological control agent escaping into the wider environment. Compared

---

26   Anon 1918, p. 1.

Figure 6.3 Prickly-pear forest 1930. (Queensland State Archives. Digital Image ID3035. Used with permission.)

to other biological control programs at the time, this program would be better termed 'controlled biology'. Its spectacular success legitimised biological control, but perhaps the spectacular eradication of prickly pear also induced a complacency that allowed for the later untested release of the cane toad. Today, the lasting results of these two programs, prickly-pear and the cane toad, stand in stark contrast. In tropical and subtropical Australia prickly-pear is reduced to occasional single plants that are hard to find; cane toads seem to be everywhere.

The success of the prickly-pear eradication program lies in the detail. The cactus had infested 240,000 square kilometres (60 million acres) of grazing land in Queensland and New South Wales; an area the size of the United Kingdom, or a little larger than the State of Minnesota. The cactus had formed impenetrable stands and made grazing impossible. There were six main species of prickly-pear cactus but the most common was *Opuntia inermis*, first released at Scone in New South Wales in around 1839 to be grown as hedges around homesteads. It had also been distributed by the Queensland Acclimatisation Society whose opinion on the unwanted spread of prickly-pear was that 'the fault is scarcely that of the Prickly pear but rather of those that permitted it to attain unmanageable dimensions and to grow it where it is not required'.[27] Quite rightly, people were the problem, not the plant.

By 1920 the cactus had become an economic disaster for graziers and it was almost impossible to control its spread. A mixture of arsenic pentoxide and sulphuric acid was the most widely used chemical control, but it had to be sprayed by men on horseback riding in amongst the cactus. 'The mixture destroyed what pear it could be sprayed on. It also destroyed the men's clothing, their boots, their saddles and eventually their horses who lost their hair and developed sores that would not heal.'[28] Mechanical destruction of the cactus was impossible because the plants simply regenerated from pieces that lay on the ground.

The only solution was biological control, using a natural predator of the plant from its native lands to destroy the infestation. As with the control of cottony cushion scale in California 30 years earlier, biological control of prickly-pear was not a lifestyle choice, it was a necessity. But by contrast with the release of the vedalia beetle in California, planning and execution of the biological control of prickly-pear in Queensland was meticulous. Work started in 1912 with a Travelling Commission of Henry Tryon and Harvey Johnson, dubbed the 'prickly pair', who

---

27   Rolls 1969, p. 355.
28   Rolls 1969, p. 355.

spent eight months visiting the home of the *Opuntia* cactus in North and South America. They recommended the establishment of the Commonwealth Prickly-pear Board 'to introduce under safeguards certain insects and diseases from America'.[29]

The innovation lay in the two words 'under safeguards'. The Board's dictum was that 'every possible safeguard should be taken to ensure that the insects introduced into Australia will not turn their attention to crops, fruit trees and other economic plants'.[30]

The Board established a research station at Uvalde in Texas where insects that fed on prickly-pear were then forced to feed exclusively on other crop plants. They had to either eat these plants or die, thereby proving they could exist only on prickly-pear. Any insects that were able to complete their life cycles on plants other than prickly-pear were excluded from the program. For the next safeguard, populations of candidate insects were bred and cosseted for generations to make sure that they were free from endemic diseases and parasites. Disease-free insects that passed this selection were then sent to Queensland to acclimatise to the seasons of the southern hemisphere. State government organisations then distributed them to areas infested with cactus and monitored their progress.

The Board tested 150 arthropods: 40 moths and caterpillars and 40 beetle grubs that attacked the stems and joints of prickly-pear; 22 cochineal and other scale insects; 15 insects that attacked flowers, fruits and seeds; 13 plant sucking bugs; 11 scavenging flies; seven gall-forming midges; and two miscellaneous insects. Just four were found suitable for release: the cochineal insect *Dactylopius tomentosus*, the plant-sucking cactus bug *Chelinidea tabulata*, the prickly-pear red spider *Tetranychus opuntiae*, and the moth *Cactoblastis cactorum*.

And the winner from a field of 150 nominations was the *Cactoblastis* moth.

This 'plain looking brown moth with a wing span of slightly over an inch [25 mm]' was found by the 'prickly pair' in Argentina in 1914, and was re-discovered there by Dodd in 1924.[31] The dowdy moth lays its eggs on cactus and its larvae, caterpillars, tunnel away inside the plant.

---

29   Dodd 1929, p. 9.
30   Dodd 1929, p. 14.
31   Dodd 1929, p. 29.

Naturally occurring rot fungi complete the destruction and 'the spectacle is afforded of the pear collapsing and dying ... as though it had been visited by some virulent plague'.[32]

The introduction of the moth to Australia was not quite as it seems. The impression given by Alan Dodd in his 1929 report on the operations of the Commonwealth Prickly-pear Board[33] was that all organisms were tested at Uvalde in Texas before being brought to Australia. However, in the case of the *Cactoblastis* moth collected in Argentina, the United States would allow the moth to neither enter nor transit through their country. As a result, moth larvae were not sent for testing to Uvalde. Instead, they were shipped from Buenos Aires via Cape Town to Melbourne, and thence to Brisbane and the Prickly-pear Board Laboratory at Sherwood. In Queensland, fresh larvae were offered 15 plants: 'fig, eucalypt, arrowroot, banana, sugarcane, peach, mango, orange, pawpaw, pineapple, cotton, maise [sic], plum, cabbage and tomato'.[34] No larvae attempted to feed, and all larvae died within 48 hours. The moth *Melitara* had been tested extensively and Alan Dodd persuaded the Board that 'for all intents and purposes *Cactoblastis cactorum* could be regarded as a species of *Melitara*'[35] and should be recommended for release. It was a convenient short-cut that stretched, but did not quite violate, the protocols established by the Commonwealth Prickly-pear Board.

The *Cactoblastis* moth was released in Queensland in early 1926. By 1928 prickly-pear was under control.

Success came from just 2,750 caterpillars brought to Australia from Argentina. To honour the moth, Queenslanders built the Boonarga Cactoblastis Memorial Hall and graziers in the Dalby district of western Queensland erected a monument to the moth in 'gratitude for deliverance from that scourge'.[36]

The spectacular success of the *Cactoblastis* moth in freeing Queensland's graziers of cactus ranks equal to the vedalia beetle's liberation of California's citrus growers from cottony cushion scale. Both the

---

32   Dodd 1929, p. 32.
33   Dodd 1929, p. 32.
34   Dodd 1926.
35   Dodd 1926.
36   Cactoblastis Memorial (n.d.).

moth and the ladybird relieved landholders of economic and psychological depression. And like the ladybird, the moth was celebrated a hero in its new land. Alan Dodd was a hero twice over. On his retirement in 1962 he was awarded an OBE.

The prickly-pear program was a triumph of selection, testing, acclimatisation and distribution. By contrast, the vedalia beetle was simply liberated with good intentions and best wishes. Unfortunately, release of the cane toad in Queensland almost a decade later would bear more resemblance to the vedalia program than to Alan Dodd's home grown strict protocols.

7
Hawai'i leads biological control

In the inter-war decades of the 1920s and 1930s, the Hawaiian Sugar Planters' Association (HSPA) was to become the world leader in biological control with an ambit far wider than just insects and sugar. Its leader, Cyril Pemberton, built on the foundations established by the HSPA's Albert Koebele, Robert Perkins and Frederick Muir. Pemberton was a rising star in the insect world and, later, was responsible for introducing the cane toad to Hawai'i from Puerto Rico and enthusiastically supporting its introduction to new locations around the Pacific. Under Cyril Pemberton's leadership, the HSPA became the epicentre of cane toad distribution.

Cyril Pemberton was a modest, meticulous and careful scientist, perhaps more so than his predecessors at HSPA. To understand how Cyril Pemberton came to promote the cane toad it is necessary to understand his work ethic, the milieu of his world of economic entomology, and how he earned his reputation as a leading entomologist of his day. Key to his success was the grower-funded wealth of the HSPA that allowed its scientists to undertake years-long journeys in Asia and Australia in pursuit of economically valuable insects. This wealth was in stark contrast to scientists at Queensland's Bureau of Sugar Experiment Stations (BSES) who, similarly tasked with protecting their State's sugar crop, rarely left home.

Figure 7.1 Cyril Pemberton, daughter Virginia and Indian motorbike 1916. (Michael A Lilly. Used with permission.)

Understanding Cyril Pemberton, the HSPA and close personal links to Australian scientists are key to understanding how the cane toad ended up in Hawai'i, South Pacific islands and Australia.

Cyril Pemberton was born in 1886 on a small citrus orchard in Los Angeles at the height of the cottony cushion scale infestations. Life was tough until Koebele's Australian ladybirds were released into surrounding citrus orchards. Citrus crops recovered and, with pressure removed from family finances, Pemberton went on to study economic entomology at Stanford University. After graduation he worked for the United States Department of Agriculture (USDA) in both California and, from 1913, in Hawai'i where he later joined the Experiment Station of the HSPA. It was a trail of coincidence. Pemberton, born into Koebele's ladybird fantasy on a Californian citrus farm, ended up, like Koebele, an economic entomologist at the HSPA Experiment Station.

Pemberton had a quiet war as a First Sergeant, supervising forest workers on Oahu, and joined the HSPA Experiment Station on the day of his discharge from the army in 1919.[1] He was 31 years old; after an amiable divorce he was a single man and a singular man, free to travel to remote jungles and swamps to search for predatory and parasitic insects to send back to Hawai'i. Frederick Muir was still at the HSPA but both Albert Koebele and Robert Perkins had left. Pemberton continued their legacy but his methods were fundamentally different. Rather than getting insects into Hawai'i and then studying them, Pemberton's philosophy was to spend most time 'in the countries from which parasites might come rather than in the countries to which they might be sent'.[2] Pemberton spent more than two-thirds of his time overseas searching remote landscapes for beneficial insects to send back to Hawai'i.

In 1920, following in Koeble's footsteps, Pemberton travelled to Fiji to collect parasites to combat the sugar cane leaf hopper in Hawai'i. While there, he received instructions to travel to Australia. His mission was to collect the tiny wasps that were responsible for fertilising the flowers of figs. The HSPA believed that some of the 17 species of fig trees that had been planted in Hawai'i would be ideal for reforesting denuded catchments on the islands. But the trees were infertile for want of wasps to fertilise the flowers. The HSPA sent entomologists to the Philippines, India and Australia in search of them.[3] The specialised life cycles of these tiny wasps is in remarkable synchrony with the life cycle of the figs – so much so that each species of fig requires its own species of wasp.

Cyril Pemberton's role in identifying, collecting, breeding and transporting Australian fig wasps back to Hawai'i is a story of skill, determination and dedication underwritten by the wealth and international influence of the HSPA and its island stewardship far broader than just the sugar industry. During his visits, Pemberton, like Koebele, Perkins and Muir before him, worked closely with the small community of Australian entomologists. In Sydney, Pemberton befriended the New South Wales government entomologist Walter Froggatt who knew a great deal about fig wasps. Indeed, the Moreton Bay fig wasp bears his

---

1   Bianchi 1977.
2   Bianchi 1977.
3   Hawaiian Sugar Planters' Association Experiment Station 1920.

name, *Pleistodontes froggatti*. They became the best of friends but this important friendship would turn very sour when, assisted by Pemberton, cane toads arrived on the Australian continent. But in the heyday of friendship, Froggatt introduced Pemberton to the Moreton Bay figs in Sydney's Botanic Gardens.

Moreton Bay figs, *Ficus macrophylla*, are barn-sized trees with spreading crowns of hand-sized, tough, dark green leaves atop short, thick buttressed and fluted stems. They start life as seed in bird droppings laid on host trees. Germinating seed sends shoots upwards and sends aerial roots down to the ground that then thicken and strangle the host. Their flowers hide inside hard leathery balls that cluster at the ends of branches. Fig flowers never see the sun but ripen to become figs after being fertilised by wasps that burrow into the tough balls.

Female fig wasps, each about three millimetres long, are born inside an unripe fig. The eggs they hatch from are laid by a previous generation of females. As female wasps hatch, male flowers inside the unripe fig shed pollen. Female wasps covered in pollen burrow out of the fig, then find and burrow into another unripe fig to mate and lay eggs. In doing so, they carry pollen from their birth fig onto flowers inside the new fig. Their work done, female wasps die and are absorbed by the ripening fig. But many females die, their short lives unfulfilled, blown away from the trees on a breeze. Male wingless fig wasps, half the size of females, never leave home. They mate with visiting females and die in figs they are born in – sedentary sperm donors. In return for fertilising its flowers, the fig gives the wasp food, water and protection from parasites and predators. Juvenile wasps are safe and secure, but the lives of adults are, as nature intended, 'singular, poor, nasty, brutish and short'.[4]

December 1920, Sydney, Australia's southern hemisphere summer. Cyril Pemberton, perched on a wooden stepladder, his head in the crown of a Moreton Bay fig, lens in hand, watched fig wasps at work. His Panama hat protected a thinning scalp and his jacket was misshapen with tools of his trade – instruments to observe and collect fig wasps. From the ladder hung a canvas bag with glass vials and a killing jar ready to receive captive wasps. In the distance was the hum of the

---

[4] Hobbes 2009(1651), p. 21.

city, a tramcar screeching the rails on the curve, its bell hustling a slow horse and dray. Cathedral bells pealed off the hours. Downslope in the harbour, ferries hooted their transits. But Pemberton was oblivious, lost in the world of fig wasps.

He focused his lens on the tip of an immature fig as a female wasp landed and searched for a way in. The tiny wasp, one quarter the width of his fingernail, found a spot at the tip of the fig and started to burrow using serrated blades on either side of its wedge shaped head. It laboured for an hour and a half. Pemberton concentrated intently, keeping the tiny wasp in focus under the lens. He plucked the fig and carefully cut it open. Inside were five females, all searching for places to lay eggs.

With tweezers, he removed a wasp and examined it. It was quite damaged. In the exertions of entering the fig it had lost its wings, antennae and most of the pollen it had carried. But it still carried enough pollen to fertilise female flowers in the fig and, in return, the fig had prepared special gall flowers to host the wasp's eggs.[5]

He was a happy entomologist. He had found the wasp and there were many more summer days in which to study its remarkable and secretive life.

Pemberton remained HSPA's 'fig merchant'[6] until late March 1921, sending batches of five species of fig and live wasp larvae back to Honolulu. Many thousands of fig wasps hatched out on the voyage and died, but those that hatched in the ship's cool room survived. As with Koebele's ladybirds, the ship's butcher was an invaluable ally.

June 1921, Emma Square, Honolulu. For the first time, figs began to ripen all over mature fig trees.[7] Pemberton's fig wasps were a success – his reputation was in the ascendancy.

During the southern hemisphere summer of 1921, relaxing between spells of collecting wasps, Cyril Pemberton spent many Sundays with Walter Froggatt's family at home in the Sydney suburb of Croydon. He wrote to Hamilton Agee, Director of the HSPA, that 'Mr Froggatt is one of my best friends here [in Sydney] and I am in close touch with him constantly'.[8] Agee asked Pemberton to remain in

---

5 Pemberton 1921c.
6 Pemberton 1921a.
7 Hawaiian Sugar Planters' Association Experiment Station 1921.

Australia and gave him another assignment. He was to collect parasites or predators of a fern weevil, the burrowing larva of a beetle, *Syagrius fulvitarsus*, that, unchallenged, was destroying tree ferns in Hawai'i. Pemberton once more enlisted the help of Froggatt who suggested that he search in the Big Scrub, an expanse of subtropical rainforest along the Richmond River in northeastern New South Wales. In unlogged rainforests near Nimbin, Pemberton soon found larvae of the fern weevil.[9] He kept them under watch until tiny parasitic wasps, *Doryctes syagrii*, hatched out of the larvae. Pemberton collected weevil larvae infested with the eggs of parasitic wasps and shipped them to Hawai'i. Back in Hawai'i, without prior testing, the wasps were released into the fern forests of Oahu and the Big Island and immediately set about decimating fern weevils. The tree ferns were saved. Cyril Pemberton's reputation was further enhanced.

Pemberton arrived back in Honolulu in early March 1922 after an absence of one year and four months – longer than he had sat in his office in the HSPA since his appointment. But he did not warm his chair for long. He soon departed for the Big Island of Hawai'i to solve the problem of rats carrying bubonic plague. He returned to Honolulu two years later with the new title of Associate Entomologist. But just six weeks later, in mid-February 1925, he left on a two year tour of the Philippines, Java, Sulawesi and north Borneo. His tasks were to find parasites of two pests of sugar cane, the mole cricket *Gryllotalpa africana*, and the sugar cane weevil borer *Rhabodscelus obscurus*. It was an arduous trip and he ended up with malaria for his troubles. In Luzon in the Philippines he found a parasitic wasp *Larra luzonensis* that preyed on mole crickets but he had no success finding parasites of sugar cane weevil borers.[10]

Pemberton was still overseas in search of wasps in 1927 when Queenslander, Arthur Bell, visited the HSPA at the end of a four year study tour. The Queensland Government's Bureau of Sugar Experiment Stations (BSES) was tasked with leading the sugar industry and the Queensland economy out of post-war depression – Arthur Bell was one of its protégés. Like medical orderly Alan Dodd, gunner Arthur Bell

---

8   Pemberton 1921a.
9   Pemberton 1921b.
10 Bianchi 1977.

was selected for a lead role on his return to Queensland. In 1924, Harry Easterby, Director of BSES, and Thomas McCawley, Queensland's Chief Justice, had arranged once-in-a-lifetime scholarships providing salaries and expenses for four years for young Queensland men to be trained overseas in the latest technologies and to cultivate them for leadership on their return – Arthur Bell was one of the lucky three selected.

Bell completed a Master of Science in plant pathology at Berkeley in California and a Diploma at the Royal College of Science at the University of London; the transition was 'like going from a palace to a cottage'.[11] In 1927, after touring southeast USA, Central America, Puerto Rico, the British and French West Indies, China, Japan, the Philippines and Java, Bell spent four months in Hawai'i at the HSPA Experiment Station.[12] This experience further embedded BSES with the scientists in HSPA.

Bell returned to Queensland in 1928 and was appointed an Investigations Officer in BSES.[13] He returned full of enthusiasm for scientific breeding of sugar cane, for rigorously designed field experiments, and for international networking with organisations such as HSPA.[14] He was also full of enthusiasm for the White Australia policy, believing that 'mentally and physically the Australian has the finest heritage in the World' and he was convinced of 'the soundness of the White Australia policy [which should be] defended to the last man'.[15] It was a belief that matched his sponsor and benefactor – the Queensland sugar industry.

Although their paths did not cross in Hawai'i, Cyril Pemberton and Arthur Bell would meet four years later when they both attended the congress of sugar technologists in Puerto Rico – where they would come face to face with the cane toad.

In the meantime Pemberton, now Chief Entomologist of HSPA, continued his scientific exploits in what seemed to be any location other than Hawai'i. In April 1928 he was invited by Elmer Brandes, Principal Pathologist in the USDA in Washington, to join his expedition to collect new varieties of sugar cane from New Guinea, the

---

11 Bell 1928.
12 Lee 1927.
13 *Queensland Government Gazette*, 28 March 1938.
14 Bell 1928.
15 Bell 1928.

ancestral home of the sweet grass.[16] Brandes chose Pemberton because he was 'a seasoned explorer who has spent much of his life seeking in remote places for parasites of sugar cane insect pests'.[17] And he chose Jacob Jeswiet, a Dutch sugar cane geneticist from Java. Richard Peck, veteran of previous New Guinea expeditions, piloted the tiny Fairchild Wasp four seater floatplane.[18] In May, the expedition staged through Sydney[19] to finalise arrangements and by June the scientists were in Port Moresby waiting for the expedition's advance party to establish a base camp on the Fly River.

June 1928, a government rest house on Paga Hill, Port Moresby, Australian administered Territory of Papua. Cyril Pemberton sat at a writing desk. Dry season afternoon breeze rustled palm fronds outside, combed kunai grass on the hill, carried resinous scent through the louvred wall. His pipe fuelled and smouldering, Cyril Pemberton penned a short letter to his colleague, Al Mangelsdorf at HSPA. The expedition was so well armed against head hunters, cannibals and crocodiles that Pemberton dubbed themselves the 'Three Musketeers'.[20] But he expressed doubts about the tiny aircraft, saying, with candid irony, 'I don't expect to live much longer, which is a pity. This seaplane that we have is certainly due to wipe out three great sugar experts within the next few months.'[21]

It was an ambitious expedition into unmapped territory. The Three Musketeers in line behind the pilot who flew the plane far into the unexplored headwaters of the Fly River, landing on remote and unmapped lakes and frightening the locals who had never seen white people, let alone a noisy aircraft. Pemberton wrote, 'if anything had happened to

---

16 The New Guinea expedition was funded by Bror Dahlberg, founder of Celotex Corporation of Louisiana that made wallboard from bagasse, the fibrous waste of sugar cane. The Louisiana sugar industry was plagued by disease, and Dahlberg, worried about future feedstock for his factories, called on the USDA to help find new disease-resistant cane varieties.
17 Brandes 1929, p. 259.
18 The floatplane was a Fairchild FC-2W Wasp, a folding wing monoplane, similar to one of the aircraft used in Rear Admiral Byrd's Antarctic expedition a year later.
19 *Sydney Morning Herald*, 4 May 1928, p. 12.
20 Pemberton 1928.
21 Pemberton 1928.

our plane ... the chances of our survival would have been very dim'.[22] Remarkably and without major mishap, the expedition travelled more than 10,000 miles (16,000 km) in the tiny floatplane, by canoe and on foot through the catchments of the Fly and Sepik rivers and along both the north and south coasts of the island of New Guinea.[23]

While Cyril Pemberton was absorbed with risks to life and limb, Elmer Brandes revelled in a cockeyed anthropology, finding 'neolithic man in unmapped nooks of sorcery and cannibalism' and coming 'face to face with naked wild man at the very dawn of reason, barely groping as yet from brute instincts and abysmal urges'.[24] Brandes treated primitive people and primitive cane varieties alike and concluded that 'races of plants, like races of people, may migrate from one far part of the world to another to multiply and replenish the earth'.[25]

In 1929, readers of *National Geographic*,[26] were absorbed by Brandes's descriptions. Today his tales of 'cruel bloodthirsty savages' and talk of the 'long hard struggle man went through on his magnificent moral climb from savagery up to civilisation' would be edited to oblivion. But like Arthur Bell's hearty endorsement of White Australia in the same year, they reflect the times and the attitudes of these educated and well travelled scientists – and of society. How different these attitudes were from those we expect today. By current standards, the expedition was deficient in even the simplest health and safety precautions, in its treatment of indigenous peoples, and in the collection and quarantine of scientific material. But understanding this difference helps us to understand how scientists of that era came to believe that the uncontrolled release of 'beneficial' organisms into the environment was 'best practice' and how, just eight years later, Pemberton himself would release cane toads on the islands.

Not content with spending six months in the wilds of New Guinea, after the expedition packed up Pemberton spent a further six months communing with insects from a small wooden hut on a rubber plantation on the Sogeri Plateau outside Port Moresby, and at Rabaul on the

---

22   Pemberton (n.d.).
23   *Sydney Morning Herald*, 20 September 1928, p. 12.
24   Brandes 1929.
25   Brandes 1929.
26   Brandes 1929.

island of New Britain. He returned to HSPA in August 1929, an absence of 16 months. On his return voyage he called in to Brisbane where he renewed his friendship with Froggatt who happened to be visiting.[27]

24 October 1929, Black Thursday. Share prices on the New York Stock Exchange plummeted. Borrowers defaulted, credit collapsed, and bankers jumped to their deaths. It was the trigger for the Great Depression.

The collapse of the stock market did not stop HSPA's star entomologist, Cyril Pemberton, from travelling in search of predators and parasites. His passport continued to be stamped by customs officers in foreign lands mildly intrigued by this lean bespectacled American. But things were much tougher in Queensland where Premier Arthur Moore struggled to keep the state solvent. He adopted

> a strategy of reduced spending, increased taxes and reduced interest rates ... abolished a number of financial grants to private industries and farmers ... increased working hours from 44 to 48 hours a week ...[reduced] the basic wage, abolished the rural award and ... removed 50% of the state's workers from the operations of the conciliation and arbitration system.[28]

It was drastic action, but it worked. Unemployment in Queensland during the Depression was less than 20% compared to 30% in southern states of Australia[29] because sugar sweetened Queensland's economy and provided a quarter of the State's revenue.

Once more, Harry Easterby, Director of BSES, used the strength of the sugar lobby to gain funds for Arthur Bell. This time it was to attend the Fourth Congress of the International Society for Sugar Cane Technologists in Puerto Rico. From Queensland, it was a three-and-a-half-month round trip by land and sea to attend a two week conference. Lobbying by Easterby began in the midst of the Depression.

Lobbying was easy. Easterby, Bell and both the Under Secretary and the Minister for the Department were all housed in the same building – the offices of the Department of Agriculture and Stock, not far

---

27   Pemberton 1929.
28   Queensland Parliament House. Member Information (n.d.).
29   Fitzgerald 1984.

from Parliament House in Brisbane. Stairs and corridors were the only impediments to a meeting. In July 1931 the Queensland Cabinet approved the trip – but without funds! Easterby called in the sugar lobby and in September 1931 Cabinet approved £200 for the trip – almost half Bell's annual salary. In the constrained times this was a spectacular success for the sugar lobby. But Cabinet made one provision: that the grant not be made public.[30] The outcry to news of an overseas junket by a public servant would have been deafening. It was Harry Easterby's final triumph as Director of BSES. Two days later, while on a field trip in north Queensland, he suffered a stroke and died. Sadly it was his last field trip before retirement.

Meanwhile, Pemberton searched the Malay Peninsula for enemies of Chinese grasshoppers, *Oxya chinensis*, that had become pests of sugar cane on Oahu. At Serdang, 20 miles south of Kuala Lumpur, Pemberton's home was a grass shack. For six months he searched meticulously for parasites of grasshopper eggs, avoided tigers and shot the heads off aggressive cobras.[31] His persistence paid off. After several weeks he found the tiny wasps that parasitised grasshopper eggs. He set up a breeding facility in the shack and when he had sufficient numbers he mailed parasitised grasshopper larvae to Honolulu.

On Oahu, the wasps took to the Chinese grasshoppers, parasitised their eggs and soon controlled their populations. It was a satisfying outcome for Pemberton, even more so because the two species of wasps he collected were new to science. HSPA taxonomist Philip Timberlake named them *Scelio serdangensis* after the place where they were collected, and *Scelio pembertoni* in honour of the collector. Among all his professional achievements, this accomplishment satisfied Pemberton more than any. Not only had he found parasites of grasshoppers where none were known before, but the parasites themselves were also unknown to science.[32] This diligence in the field was Pemberton's hallmark but it was missing when later, rushing to despatch cane toads to Hawai'i, parasites were shipped unintentionally together with the toads.

In his first 12 years working for HSPA, Pemberton greatly increased populations of exotic insects on the Hawaiian Islands. Not all of

---

30   Kerr 1931.
31   Pemberton (n.d.).
32   Bianchi 1977.

them were successful. To control pests of sugar cane, he introduced the mirid bug *Cyrtorhinus mundulus* from Fiji that controlled sugar cane leaf hopper *Perkinsiella saccharicid*, the larrid wasp *Larra luzonensis* from the Philippines that controlled the mole cricket *Gryllotalpa africana*, and two wasps *Scelio serdangensis* and *Scelio pembertoni* from the Malay Peninsula that controlled the Chinese grasshopper *Oxya chinensis*. But an unnamed fungus from Java and another he introduced from Sulawesi failed to control the sugar cane weevil borer *Rhabdoscelus obscurus*.

Working outside sugar cane, he introduced the fig wasp *Pleistodontes frogattii* from Australia that fertilised Moreton Bay fig trees, the braconid wasp *Dorcytes syagrii* from Australia that controlled the fern weevil *Syagrius fulvitarsus*, and the moth *Bactra truculenta* and weevil *Athesapeuta* from the Philippines that partially controlled nut grass *Cyperus rotundus*. But an unnamed fly larva from southern Sulawesi and an unnamed nematode from north Borneo failed to control termites, and the cannibalistic mosquito *Toxorhynchites inornatus* from Rabaul in New Guinea failed to control other mosquitoes in Hawai'i.

These are impressive achievements for one man working alone, without communications, in remote tropical jungles, a long sea journey from laboratories and the supporting infrastructure of home. But perhaps more startling from today's perspective is the ease with which Pemberton and his HSPA colleagues released exotic insects into the environment of the Hawaiian Islands. At the time it was unremarkable because HSPA, the wealthy, influential and most economically important agricultural research station on the islands, was the sponsor of the introductions. This helps to explain why, later, the cane toad, championed by HSPA, was introduced and released so easily.

No new pests of sugar cane had appeared in Hawai'i since 1917, a fact that Pemberton attributed to observance of rigid quarantine regulations governing importation of soils and regulations covering the importation of new cane varieties.[33] But in 1932 one pest still worried Pemberton: it was the anomala beetle *Anomala orientalis*, a scarab beetle whose larvae, white grubs, ate the roots of sugar cane. It was parasitised by the scoliid wasp *Scolia manilae* introduced to the islands

---

33   Pemberton 1932e.

from the Philippines in 1916, but the relationship between host and parasite populations was unstable. For Pemberton, effective control of anomala was 'a major problem confronting the entomologists'.[34]

In February 1932 both Cyril Pemberton of HSPA and Arthur Bell of BSES headed to Puerto Rico and the Fourth Congress of the International Society of Sugar Cane Technologists. Bell travelled on the regular slow boat from New York, the 11,000 ton *Borinquen*, 'pride of the Porto Rico Line'. Pemberton and delegates from HSPA travelled 'by swift modern passenger planes'.[35] BSES on a slow boat, HSPA on a seaplane clipper – a world of difference.

Bell's seven week trip began with a voyage to San Francisco via Honolulu, then rail to New York and a final voyage to San Juan, Puerto Rico. At the beginning of February 1932, he entered a California of the dustbowl and the Depression – very different from the 1920s when he wandered the marbled halls of Berkeley. In wintry New York, Bell read of a mother and daughter, both jobless, who gassed themselves and left a note to say they preferred 'death to charity'.[36] In the midst of economic woes, Russeks on Fifth Avenue was clearing out fur coats 'at cost ... below cost ... regardless of cost'[37] while destitute families on the street outside relied on the Clothing Relief Committee to keep out the winter cold. Bell left a wintry New York behind and sailed south on the SS *Borinquen* to the familiar warmth of the tropics.

Pemberton was accompanied on his flight to Puerto Rico by John Waldron, Chairman of HSPA and Director of Oahu Sugar Company. Waldron was very concerned about the damage anomala beetles were doing to his cane fields on Oahu and he had the ear of his Chief Entomologist for the next few weeks. In Puerto Rico, both Pemberton and John Waldron would come to know of another organism that might combat anomala beetles. It was an anuran, the giant tropical toad *Bufo marinus*, and another order of animalia to add to the list of Hawaiian introductions.

---

34   Pemberton 1932e.
35   Pemberton 1932b, p. 10.
36   *The New York Times*, 22 February 1932.
37   *The New York Times*, 22 February 1932.

Old friends and new acquaintances assembled in San Juan, Puerto Rico, on Monday 29 February 1932 – a Leap Year. It would prove to be a good year for toads.

8
Birth of a myth

March 1932. The people of San Juan Bautista de Puerto Rico were proud to host the sugar technologists, proud of their crops and their industry and proud of what they had achieved. None more so than the City Manager who welcomed the 'missionaries of science' and gave them the freedom of the city. They would feast and be feted on sugar estates and in sugar mills as they journeyed in convoys of cars around the rim of the island. A select few would hear extraordinary stories of how *Bufo marinus* saved the island's sugar industry, a creature so new to the island that it had no local name except bufo.

Puerto Rico is a rectangular-shaped island in the northern Caribbean. Its central cordillera clothed in rainforests captures seasonal rains and its rivers provide irrigation for sugar estates on dry fringing coastal plains. Sugar was the island's economic mainstay. Like Hawai'i, in the three decades since the United States had taken control, the island of Puerto Rico had been treated as another huge experiment in biological control.

In 1901, US Congress authorised establishment of the Mayagüez Agricultural Experiment Station and the USDA provided expert staff. Results were spectacular. Sugar cane production doubled within 10 years and trebled in the following decade. But as sugar production increased, so did the problems. In the drier southwest of the island, fields of mature cane yellowed and withered; sick plants had no roots. In some years, the problem was so severe that cane was imported from the Dominican Republic to maintain mill capacity and exports.[1]

The cause was white grubs, the soil-dwelling larvae of large native scarab beetles, called May beetles.[2] Like cane grubs in both Hawai'i and Queensland, Puerto Rico's native white grubs enjoyed the sweet roots of sugar cane. They were the pest of every crop grown[3] and 'in bulk, the white grubs of Puerto Rico greatly overbalanced all the other insects of the island'.[4] The problem had increased steadily. What caused the increase in populations of May beetles and white grubs?

Some blamed the Indian mongoose *Herpestes birmanicus* introduced to Puerto Rico between 1877 and 1879.[5] The mongoose was widely used in the Caribbean to control rats in sugar cane, but it created havoc. In Puerto Rico it ate its original target, the ground-dwelling Norway rat, but this caused an increase in populations of a bigger pest, the roof rat. Mongoose also host rabies and, just like rats, carry *Leptospira* bacteria.[6] The bacteria are transmitted in the animal's urine and cause Weil's disease in humans; flu-like symptoms that that can lead to meningitis. It is also known as canefield fever.

In Puerto Rico, the mongoose ate ground-nesting birds, snakes, lizards and domesticated chickens. Most importantly for cane growers, the mongoose wiped out the Puerto Rican iguana *Ameiva exsul* – a ground-dwelling whiptail lizard. The lizard spent only a little time above ground around midday and spent the rest of the day burrowing through the soil eating white grubs.[7]

Some blamed exotic trees for the white grub problem. Two Australian trees, *Casuarina equisetifolia* and Silky oak, *Grevillea robusta*, had been introduced to Puerto Rico as shade trees and adult May beetles were seen to favour these for resting after their flight from the cane fields. For some farmers that was evidence enough to cut down all the foreign trees.

---

1 Wolcott 1950, p. 188.
2 *Phyllophaga vandinei* and *Phyllophaga portoricensis*.
3 Wolcott 1948, p. 250.
4 Wolcott 1935, p. 447.
5 Wolcott 1950, p. 185.
6 Howarth, 1991, p. 498.
7 Wolcott 1950, p. 186.

But there was another, simpler explanation that the white grub problem was due to the increased area of sugar cane plantations. And 'for practical purposes … white grubs had no natural enemies of importance on the south coast of Puerto Rico'.[8]

The most important question was how to control white grubs. Chemicals were ineffective. Paradichlorobenzene, an ingredient of mothballs, was poured onto the soil – cane plants died but white grubs survived. Undiluted carbon bisulphide, a fumigant insecticide, was injected into the soil around cane plants, but again white grubs survived.

Some tried explosives:[9] sticks of dynamite laid out in a regular grid pattern, wired and detonated. The explosions were impressive. Soil and grubs were hurled high in the air and the landscape resembled a scene from the Western Front. But after they recovered from the blast, white grubs burrowed back into the soil once more to find some peace and quiet. And what to do with the craters?

At least one scientist seriously considered importing the North American skunk to chase grubs[10] but colleagues dissuaded him.

Birds helped. Native blackbirds and herons followed the plough and ate white grubs, but they could only eat so much. So farmers supplemented them with pigs that rooted through the soil, searching out grubs until they laid down for a snooze. On the other hand, gangs of women and children could collect white grubs, did not lie down if the foreman was watching, and did not have to be fed and housed. On one property, gangs of women and children collected two tons of grubs.[11]

Cane farmers were frustrated. There appeared to be nothing they could do to reduce populations of white grubs, so they took matters into their own hands. In 1910, the Puerto Rico Sugar Producers' Association funded their own Experiment Station on the north coast, at Rio Piedras. Building laboratories had worked wonders for cane growers in the HSPA in Hawai'i so it should do the same for the industry on the north coast of Puerto Rico. They just had to hire the scientists. And science would find a natural predator of white grubs – a biological control.

---

8   Wolcott 1950, p. 188.
9   Wolcott 1948, p. 260.
10 Wolcott 1950, p. 189.
11 Wolcott 1935, p. 454.

An Experiment Station would solve the cane growers' problems. It was a matter of faith.

At Rio Piedras the scientists' first choice for the tropical island of Puerto Rico was a scoliid wasp, a species of *Tiphia* from the sometimes baking hot and sometimes snowy plains of Illinois. In the middle of the American Prairie, scoliid wasp larvae sometimes have to tolerate temperatures of -18°C and a metre of snow. The scoliid wasp, about 25 mm or an inch in length, hunts white grubs in the soil, paralyses them with a sting and then lays eggs on them. When wasp eggs hatch, wasp larvae eat white grubs.

The wasp migration program was approved and funded by the Puerto Rico growers. A 'Travelling Entomologist' was appointed to collect wasp cocoons from Illinois and send them to Puerto Rico. But he was killed when his motorbike was struck by an inter-urban electric railroad car. His misfortune became the good fortune of George Wolcott, hired to replace him. Wolcott, from Utica in upstate New York, took over the task of collecting wasp cocoons in Illinois. Forsaking the motorbike, Wolcott made his collections using the very railroad that had caused his predecessor's demise.[12] He dutifully collected and despatched wasp cocoons for six years. When each batch of wasp cocoons arrived in Puerto Rico they were placed in cane fields infested with white grubs. But at the end of each year, and after six years in a row, no Illinois wasp cocoons could be found in the soil, no white grubs hosted wasp larvae, and no Illinois scoliid wasps flew in the Puerto Rican cane fields. There was nothing to show for six years of work. The wasps had failed.

With a scientist's customary caution and capturing the essence of six years of research, Wolcott crafted the hypothesis that 'the fundamental difference in climate between Illinois and Puerto Rico is so great that it seems highly improbable that such a specialised insect could survive the change'.[13] It was a somewhat obvious hypothesis.

The Illinois wasp had no impact at all apart from keeping Wolcott employed. In any case, Puerto Rico had its own scoliid wasps that did not control white grubs either because the wasps were parasitised by local bee flies – called bombyliid flies. Undaunted, Wolcott carried on,

---

12   Wolcott 1935, p. 447.
13   Wolcott 1935, p. 448.

8 Birth of a myth

importing scoliid wasps from Barbados and from neighbouring Haiti, including one species of wasp that fed on the pollen in flowers of wild parsnip. The team at Rio Piedras worked hard at growing wild parsnip but they could not get it to flower at low altitude on the dry plains where cane was grown, so that plan was also abandoned.

The next idea was to use the native soil-dwelling larvae of the luminous cucubano beetle *Pyrophorus luminosus* that voraciously ate any soil-dwelling larvae. They were collected from the rainy central cordillera of Puerto Rico and transported to the cane fields of the dry northern coast – where they died.

Frederick Muir at HSPA had success with tachinid flies as predators of beetles. On Hawai'i, they would attack adult beetles in flight and lay their eggs in the soft parts of beetles' abdomens. Pursuing this lead, the scientists at Rio Piedras collected tachinid flies from the rainy central cordillera and took them to the cane fields of the dry northern rim – where they died as well.

Scientists at the industry-funded Experiment Station at Rio Piedras failed to find a biological control for white grubs in Puerto Rico. So the sugar industry stopped their funds. In 1913 the private owners handed the Station over to the Government of Puerto Rico to be run on public funds. It was renamed the Insular Experiment Station.

George Wolcott enlisted when America declared war in 1917. After returning from the Western Front, he did not settle and moved around the Caribbean and South America before returning once more to Rio Piedras.[14]

After all the failure came a saviour in the form of the toad.

In 1920, without fanfare, David May of the USDA, Director of the Mayagüez Experiment Station, brought a dozen 'Surinam toads', *Bufo marinus*, from Barbados and released them in cane fields in the southwest of Puerto Rico.[15] But, when questioned later by Wolcott about his role in introducing toads, he was 'singularly diffident about admitting his complicity [in their release]'.[16] The toads liked the irrigated cane fields and started breeding.

---

14 Lawrence 2000.
15 Tucker & Wolcott 1935.
16 Wolcott 1950, p. 189.

Figure 8.1 Bufo marinus pair in amplexus. (David Nelson, Desert Ecology Research Group, School of Biological Sciences, University of Sydney. Used with permission.)

In 1923 the Insular Experiment Station at Rio Piedras followed suit. Menendez Ramos, the station's Director, was in Kingston, Jamaica, waiting for a boat back home. Toads ringed the circle of light cast by a solitary lamp and Ramos saw 'the eagerness with which they snapped up any May beetle or other large insects that happened to alight near them'.[17] Locals told him the big toads ate everything, even rats! That was the extent of this scientist's investigation of *Bufo marinus*. It was evidence enough. Ramos took 40 toads back to Puerto Rico and released them at Rio Piedras.

Within 10 years of its introduction, *Bufo marinus* had spread throughout the cane fields of Puerto Rico. Wolcott was impressed with its powers. In one cane field 'cane leaves had been eaten to the midribs by the beetles'. Toads were collected and released and 'a year later in the same field hardly any indication of the presence of beetles could be found'.[18] Wolcott reasoned that toads, when released, had eaten the

---

17 Wolcott 1935.

adult beetles because there was no further browsing of leaves. And if the beetles had been eaten, there would be no white grubs either. No further evidence was needed. No need to confirm the hypothesis by digging for grubs or trapping flying beetles. Wolcott's logic was medieval. Toads appeared and beetles vanished, 'for thus evil takes away evil'.[19] His conclusion was closer to witchcraft than science.

Nevertheless, George Wolcott anointed *Bufo marinus*: 'This is one of the few instances on record of a foreign predator being entirely successful in the control of a native insect pest and, to date, the toad has developed no serious undesirable habits'.[20] With no supporting evidence, Wolcott declared that the habits of the May beetles simply 'dovetail perfectly with the food requirements of Bufo'[21] and 'the solution of the white grub problem by the introduction of the toad was indeed a major triumph for biological control'.[22]

After years of failure at Rio Piedras, the toad was hailed as the saviour of cane growers and it also defined the worth of the scientists in the Experiment Station at Rio Piedras. It was their longed for success. Glowing with euphoria, no one was inclined to question Wolcott's claims.

Further 'scientific' proof arrived a little later. Raquel Dexter, MSc from Chicago, biology teacher in the Colegio de Agricultura y Artes Mecanicas (the College of Agricultural and Mechanical Arts) at Mayagüez, investigated the feeding habits of *Bufo marinus*. She followed the example of Archie Kirkland,[23] a USDA entomologist who studied the feeding habits of the American toad, *Bufo lentiginosus americanus*, that eats May beetles in its homeland. Kirkland dissected 149 toads and drew up a 'balance sheet' of beneficial, neutral and injurious insects found in their stomachs. He believed his results showed 'the [American] toad to be a highly beneficial animal and well entitled to man's protection in every possible way'.[24]

---

18   Wolcott 1935, p. 455.
19   Leeser 1959.
20   Tucker & Wolcott 1935, p. 399.
21   Wolcott 1948, p. 251.
22   Wolcott 1948, p. 261.
23   Kirkland 1904.
24   Kirkland 1904, p. 13.

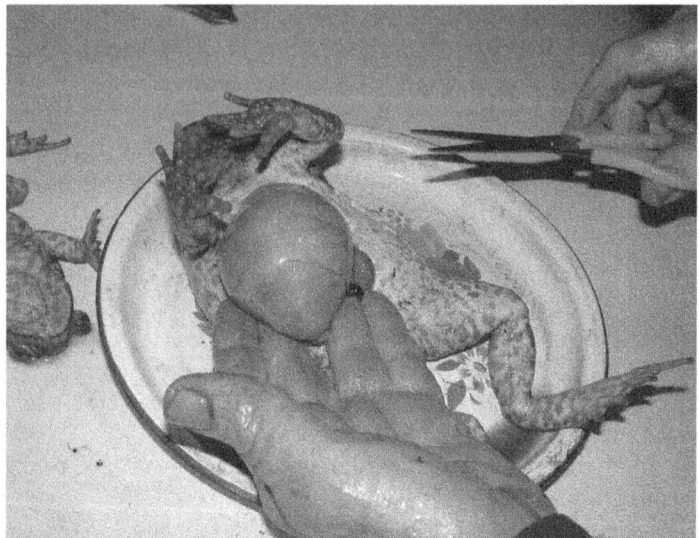

Figure 8.2 Dissection of the stomach of Bufo marinus. Bulburin, Queensland. (Kipling Will. Reproduced with permission.)

With Archie Kirkland's methodology in hand, helped by Puerto Rican cane farmers and her colleagues Mortimer Leonard and George Wolcott, Raquel Dexter collected 301 toads and cut open their stomachs. Inside the toads' stomachs the insects they have eaten are remarkably well preserved and quite easy to identify – they are not masticated and simply dissolve over time in stomach acids. Dexter estimated that 7% of the insects in toads' guts were beneficial to man, 42% were neutral and 51% were agricultural pests[25] – results similar to Kirkland's.

Here was proof of the toad's powers. Dexter triumphantly presented this proof to the audience of international scientists at the ISSCT Congress in Puerto Rico.

Late evening, Friday 4 March 1932, San Juan, Puerto Rico. Seventeen tired Congress delegates, still recovering from the Governor's garden party, smoking and chatting, were called to order by the Chair-

---

25 Dexter 1932, p. 4.

man, HSPA entomologist Cyril Pemberton. Mrs Raquel Dexter stepped forward to be introduced. Master of Science from The University of Chicago,[26] returned home[27] to teach biology in the Colegio de Agricultura y Artes Mecanicas de Mayagüez, here presenting her paper on 'The food habits of the imported toad Bufo marinus in the sugar cane sections of Porto Rico'.[28]

The paper went well. Dexter concluded triumphantly that the toad 'can be effectively used as a biological control of *Phyllophaga* [May beetles]'[29] and she praised the 'amphibian immigrant which is doing its full share of benefit to our sugar industry, and to which this International Congress should pay a tribute of gratitude'.[30]

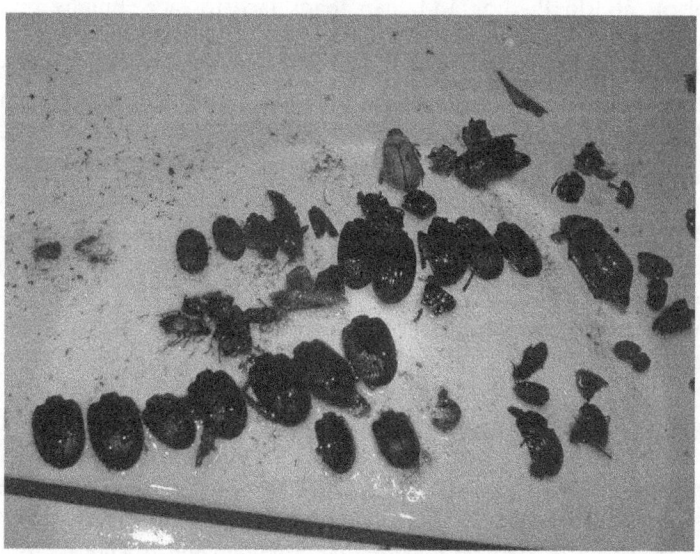

Figure 8.3 Stomach contents of Bufo marinus. Bulburin, Queensland. (Kipling Will. Reproduced with permission.)

---

26   Dexter 1929b.
27   Dexter 1929a.
28   Dexter 1932.
29   Dexter 1932, p. 4.
30   Dexter 1932, p. 5.

Polite applause and a discussion followed. Harold Box from Antigua in the West Indies complained that toads did not control *Phyllophaga* beetles on Antigua, but were a nuisance and ate bees. Cyril Pemberton, ever the gentleman, nobly deflected criticism from the lady presenter. He remarked that there were 'still in the Hawaiian cane fields a few insects not yet under complete control … [and] the introduction of toads offered possibilities for improving this state of affairs'.[31] It was a perfect segue to Pemberton's own paper, the next to be delivered, 'Recent introductions of insects beneficial to the sugar cane industry of Hawaii'.

BSES's Arthur Bell had joined the Congress field trip. Travelling around the island in a car with Pemberton,[32] they had likely met George Wolcott at Rio Piedras and seen toads face to face. But he missed Raquel Dexter's presentation on the feeding habits of *Bufo marinus* because he was busy delivering five papers from BSES scientists elsewhere in the Congress's concurrent technical sessions. Bell would not see Dexter's paper until the Congress proceedings were published a year later.

Pemberton's candidates for the toad's attention in Hawai'i were anomala grubs, 'a notorious foe to sugar cane'.[33] Like Puerto Rican May beetles they are scarabids and, since their introduction to Hawai'i in 1908, their grubs had 'developed to injurious numbers in spots of considerable size'.[34] In 1915, Frederick Muir of the HSPA introduced a scoliid wasp from Japan that largely controlled populations of anomala beetle.[35] But the pest reappeared in 1930 and caused extensive damage[36] to sugar cane on red soils at the Waipahu property of the Oahu Sugar Company 'in spite of a great abundance of the [wasp] parasite *Scolia manilae*'.[37] Moreover, John Waldron, Pemberton's boss, travelling companion and President of HSPA, was also a Director of Oahu Sugar Company and was particularly interested in solving the anomala problem. Waldron insisted Pemberton take *Bufo marinus* to Hawai'i for this

---

31 Pemberton 1932b, p. 138.
32 Pemberton 1932f.
33 Illingworth & Dodd 1921, p. 68.
34 Pemberton 1932e.
35 *Anomala orientalis*.
36 Pemberton 1964.
37 Pemberton 1932a.

8 Birth of a myth

purpose.[38] Approaching his 60th year, John William Waldron, English born, powerful businessman, Freemason, was to be disregarded at one's peril.

Pemberton, entomologist, methodical scientist, stayed on in Puerto Rico to discover more about the toad and to make sure it would not become a pest on Hawai'i. Others who helped him to a decision were Major Chapman Grant, 'an authority on reptiles ... a very solid, conservative individual',[39] Dr Mortimer Leonard, Dexter's mentor at Mayagüez, and Mrs Dexter herself.

Six days after the end of the Congress, Pemberton took tea with Dexter.[40] It was the catalyst: for Pemberton, the quiet conversation with the island's leading proponent of the toad tipped the scales in favour of the toad. Though still a little hesitant, the following morning he wrote to Otto Swezey, his colleague at HSPA:

> I have about decided to introduce to Hawaii the Central American ... toad Bufo marinus ... I wish you and [FX] Williams and Van [RH Van Zwaluwenburg] would think this over and see if you can decide on any strong objections to its introduction to Hawaii. Mr Waldron is for it 100 percent.[41]

Pemberton, still uncomfortable with the urging of Waldron, debated the case with Swezey over four pages of manuscript, anticipated his friend's questions, answered his own doubts, inquired about import permits, and concluded that the toad 'would do more good than harm if established there [in Hawai'i]'.[42] It was a weighty subject, a decision not taken lightly, but it was made. Pemberton went toad mustering.

Monday 28 March 1932, Pemberton, Mortimer Leonard and his assistant Francisco Sein collected toads at the Rio Piedras Experiment Station. The next day, Pemberton delivered 34 toads to the Porto Rican Express Company for consignment to Honolulu via New York at a cost of $10.[43] Pemberton was wary of shipping other pests to Honolulu

---

38   Pemberton (n.d.)
39   Pemberton 1932c.
40   Pemberton 1932g.
41   Pemberton 1932c.
42   Pemberton 1932c.

along with the toads, so he packed them in 'dry clean excelsior [wood wool] taken in a grocery store from a New York packing case' and warned Swezey to burn both the box and the excelsior after removing the toads. As an afterthought he cabled Swezey to check the toads for ticks that might have burrowed into the toads' skin.[44] Later, he wrote, 'This [first] shipment is quite a gamble but well worth a trial … everyone in Porto Rico, Santo Domingo and Vieques Island have nothing but praise for this toad … it has, on paper, much in its favor for us'.[45]

Pemberton held a qualified enthusiasm for the toad – that the toad was not harmful and probably beneficial. In Puerto Rico no large predator seemed to attack the toad, there were no reported deaths and no mention of its toxicity.

On Thursday 31 March 1932, toads were on their way. In the Leap Year it was the Great Leap Forward for *Bufo marinus*. With this first shipment, the genus *Bufo* would leave the Caribbean to become established in the Pacific – something 40 million years of anuran evolution, 80 million years of volcanic activity in the Pacific Ocean, and 450 million years of continental drift had been unable to do. It was the greatest expansion of the species' range since its ancestor, Fossil 41159, hopped along the flood plain of the Magdalena River on the northern coast of Colombia 15 million years before.

Pemberton sent a total of four consignments of toads to Honolulu. After the initial shipment, he sent two more consignments of 36 and 60 toads each through the Porto Rican Express Company via New York. Then he took the final consignment with him to Los Angeles. For this shipment, Pemberton purchased two suitcases, cut slots in them, packed them with moistened wood shavings, and packed them with 36 toads collected from the Forestry Station in San Juan.[46] Two days later, he and the suitcases boarded the clipper seaplane for Miami for a 12 hour island-hopping flight to Jacksonville, then New Orleans, before finally landing at Los Angeles. This 75 hour trip was the fastest passenger service available in 1932. The toads were fine. Primitive amphibians were unruffled by modern air travel.

---

43   Porto Rican Express Company 1932.
44   Pemberton 1932d.
45   Pemberton 1932c.
46   Jarvis 1934.

8 Birth of a myth

Thursday 21 April 1932, Los Angeles waterfront. Pemberton tended his charges, moistened their box for the long sea voyage and bade them farewell. Just before 9am Pemberton 'placed 36 *Bufo marinus* on the SS *City of Los Angeles* in the purser's care';[47] it was sailing the next day for Honolulu. After carrying toads from the Caribbean to the west coast of North America, island hopping in piston-driven aeroplanes, conquering continental drift and altering anuran biogeography for eternity, Cyril Pemberton had a rest. As he went on leave to Banning, California, to catch up with his immediate close family,[48] his bufos passed Angel's Gate Lighthouse and headed for their new home in Hawai'i.

But there was something amiss. Something was not quite right despite all the praise for the toad. The problem was that Dexter's conclusions could not be substantiated by her results. From her experiment it was impossible to say that *Bufo marinus* controlled May beetles as she claimed. Like Wolcott's conclusions before her, Dexter's conclusion was a leap of faith, wishful thinking, a myth – not science.

There were two failures of scientific investigation. Firstly, no one tried to find out why white grubs had become pests in cane fields in Puerto Rico. Possible causes were the mongoose, exotic trees, favourable weather (for grubs) and expanded cane growing areas. The answer to why their populations had increased to pest proportions would have provided some indications as to their population dynamics. It would also have contributed a likely cause for their decline in numbers, synchronous with the introduction of toads. But this was not a fatal flaw.

Secondly, Dexter demonstrated that *Bufo marinus* ate May beetles – and pretty well anything else. But this did not prove that toads controlled May beetles, nor did it prove that they controlled white grubs, the larvae of May beetles and the real pest. Dexter had no measure of the populations of May beetles and no proof that toads had any significant impact on populations of these beetles, or any other insect. Proof of control required a different experimental design, one that measured the dynamics of environmental factors and populations of beetles and toads. Dexter concluded that just because toads ate beetles they

---

47   Pemberton 1932c.
48   Pemberton 1932h.

controlled their populations. That was simplistic thinking, failed logic, flawed science.

These failures are simple and easy to understand. However, Pemberton was a senior scientist in a respected institution, an entomologist with a pedigree, Stanford-trained with almost 20 years experience in Hawai'i. How could these flaws have been overlooked? During the course of the Congress, Pemberton had been pressured by his boss, Waldron, to introduce toads to Oahu. But that does not explain why neither Pemberton nor other delegates pointed out the flaws in Dexter's conclusions. At the Puerto Rico Congress, only Harold Box, sugar scientist from Antigua, had spoken up to doubt the toad, but not Dexter's conclusions. It is a conundrum but one that, sadly, had sombre implications for the world outside the Caribbean.

A simple question: was the disappearance of May beetles due to the weather? In 1984, Bill Freeland in Darwin, Australia, watching the toad front advance westwards across the continent, did a simple check of rainfall in Puerto Rico between 1931 and 1936 when toads were supposedly working miracles there. Not only were these years unusually wet, but wet seasons were wetter and dry seasons drier than average,[49] likely pinching white grub breeding from either end and likely causing their temporary decline. No toads were required.

In Puerto Rico's cane fields, populations of cane toads declined despite an increase in populations of white grubs – their surmised food source. In 1951, George Wolcott still believed that 'the giant Surinam toad, *Bufo marinus* L., proved to be the solution, at least temporarily, of the white grub problem [in Puerto Rico], absolutely perfect for the ten or fifteen years when toads were most abundant'.[50]

Raquel Dexter spawned a myth, not scientific proof, but with no one to point out obvious flaws, the myth flourished. And based on that myth, Cyril Pemberton's four consignments of toads were on their way to the Pacific paradise of Hawai'i.

---

49   Freeland 1984.
50   Wolcott 1951, p. 13.

9
Toad fantasy

All four of Cyril Pemberton's bufo shipments arrived in Honolulu during April 1932. Some had been in transit for around 20 days without food or water but were still in 'good condition' when they arrived – only five dead across the four shipments. For the meticulous Pemberton, a small question of numbers: he reported 154 toads consigned, shipping dockets totalled 166 toads,[1] five were dead on route, thereby leaving depending on the report 161, or 149,[2] or 148[3] bufos to be liberated in their new home. But this detail mattered little because the import permit was open ended. David Fullaway of HSPA had negotiated 'full approval of the Board of Agriculture for the introduction of the toads'.[4]

Monday 18 April 1932, in Pemberton's absence – not an unusual situation – Fred Denison, HSPA Agriculturalist for Oahu, released bufos in the arboretum[5] near the head of the Manoa Valley, behind Honolulu.[6]

Friday 22 April 1932. Denison released bufos in a taro patch bordering a rice field near the HSPA research station at Waipio[7] to the west

---

1   Porto Rican Express Company 1932.
2   Pemberton 1933.
3   Pemberton 1934.
4   Swezey 1932.
5   Harold L Lyon Arboretum. University of Hawaiʻi (n.d.).
6   Pemberton 1933.

of Honolulu on land leased from the Oahu Sugar Company amid plantations infested with anomala beetles.

Friday 29 April 1932. Denison released another batch of bufos in the arboretum.

Saturday 30 April 1932. Denison released more toads into the taro patch and rice field at Waipio. John Waldron, HSPA President, Director of Oahu Sugar Company, would have been acutely interested in their reproduction and progress into his sugar plantation.

The prestigious HSPA had started a new trend. Once word got around the sugar-growing world that the toad had been recruited for action by the HSPA, it became the latest technology, the best in biological control and the creature of choice for cane growers. Just 40 years after Koebele's ladybird fantasy swept the orchards of the world, it was the turn of the toad fantasy to become the fashion of agriculture.

On Thursday 9 June 1932, the SS *Monterey* of the Matson Line docked in Honolulu. Pemberton returned home from Puerto Rico, back among islanders, the soft aloha, gentle protocols, customs of courtesy. He checked in to the Pleasanton Hotel where, for the previous four years, he had been a regular guest when in Honolulu. The hotel, a spreading two story plantation house surrounded by four acres of palms and tropical gardens, graced the corner of Wilder and Punahou Streets just a city block from HSPA offices. At 46 years old and despite years of hunting insects in remote tropical jungles, Pemberton was fit and healthy, tall, brown eyed with a widow's peak of close cropped hair, lean muscled and trim from daily swimming and surfing whenever he could. His closest companions were a Panama hat, round rimmed glasses, a pipe and tobacco, and a folding nest of magnifying lenses in case an interesting insect appeared. Pemberton went to his office to catch up with colleagues, then a swim in the surf at Waikiki and a spin on his Indian motorbike. He was renewed.

Bufos were left alone for a year to breed in the relative wilds of Manoa and Waipio.

At HSPA laboratories in Honolulu, toads were kept in cages with access to water and fed a diet of ants for more than a year, but failed to breed in captivity. Pemberton, now Executive Entomologist, decided

---

7   Pemberton 1934.

9 Toad fantasy

to test the range of what they would eat. Cockroaches and slaters were easily dealt with. Caterpillars and earthworms provided no challenge except that both hands were used 'in a very human but inelegant manner' to get the whole worm into the mouth. Centipedes grasped the toads' faces with their legs, and toads had to use both arms to get the meal inside their mouths. Beetles were swallowed whole but it was a struggle for a toad to get a whole Chinese grasshopper, legs and all, down its throat. A gecko was dropped in but it moved too quickly to be caught until Pemberton picked up the 'tender reptile' and dangled it in front of a toad that swallowed it in one gulp. Snails in their shells, despite their size, went down a treat. The entomologists then tried heavy artillery: spiders were eaten with a 'lightning-like flash of the toad's tongue': scorpions, their tails up and striking, were swallowed without difficulty: toads' tongues picked up wasps when they got in range: and after swallowing such a fiery creature as a de-winged carpenter bee, the toad was observed to 'execute a few abdominal motions suggestive of the Hawaiian hula dance'.[8]

With toad populations static in the laboratory, Fred Denison set up a breeding facility at Waipio. He collected gelatinous strings of fertilised eggs from the rice fields. He placed the eggs in large shallow pans of running water set into the ground under mango trees and fenced around with mesh. As tadpoles hatched and left the pans, he fed them things he ate himself like poi (fermented taro), boiled rice, cooked cream of wheat and carnation milk flakes. They grew faster and matured more quickly on Denison's diet than on ants and occasional large insects.

From August to December 1933, small numbers of toads were sent to cane fields of the Oahu Sugar Company[9] – John Waldron was one of its Directors. Wider distribution of young toads to sugar planters began in January 1934.

Cyril Pemberton rode out to Waipio to see the progress. He dressed for the field – a white shirt and tie, khaki jodhpurs, leather boots, buckled leather gaiters over his calves, and his familiar jacket, with its pockets for lenses, notebook, pipe and tobacco. He kick started the In-

---

8   Pemberton 1934.
9   Hawaiian Sugar Planters' Association 1933.

dian motorbike and roared off down the track towards Waipio, 15 miles (24 km) to the west of Honolulu.

The white, crushed coral track to Waipio crossed the grain of the country. It rose to crest gentle slopes that climbed away on his right up to knife-edged ridges with their peaks still in morning cloud. The track fell again to cross meandering creeks that ran out on his left through tidal flats and mangroves into Pearl Harbour. The air warm and moist, a million mirrors on the blue waters. Rolling country patterned green, cane fields taller than a man, ordered and even, hemmed the track on both sides begrudging it passage.

Soon the tyres were crunching red friable volcanic soil at Waipio, stirring red dust. Pemberton hauled the Indian to a stop, dismounted and stamped the dust from his boots. He dropped his goggles leaving white imprints in red ochre. Red creases bracketed a smile as Denison approached, hand extended, aloha for the boss. Then the rainwater tank, a refreshing drink and handfulls over the face and neck, red dust turning to pink stains, the distinctive branding of the cane fields.

Pemberton was impressed. The ground around the pans was alive with toads, the pens overcrowded; it was time to distribute the new anurans. Pemberton's plan was for toads to be sent to cane fields in all corners of the Hawaiian Islands.

By September 1934, Pemberton and Denison had distributed at least 1,000 toads to every sugar plantation in the islands, a total of 200,022 toads barely satisfying the demand from cane farmers – a large number had gone to anomala-infested fields of the Oahu Sugar Company. Toads had been displayed at the Kauai County Fair, Kauai High School and the Hawai'i Food Products Show.[10] The public loved them, for the most part: 'much benefit is often reported by residents … although an occasionally irate citizen objects to toads trespassing on his premises'.[11] A year later, toads were well established in many of the islands and HSPA ceased their distribution, except for special requests.[12]

Pemberton believed that 'importation to other parts of the Pacific will naturally follow'.[13]

---

10 Hawaiian Sugar Planters' Association 1934b, p. 22.
11 Hawaiian Sugar Planters' Association 1936, p. 22.
12 Hawaiian Sugar Planters' Association 1935c.
13 Pemberton 1934.

9 Toad fantasy

Figure 9.1 Cyril Pemberton, HSPA entomologist with Bufo marinus. (Bishop Museum [SP_204773]. Used with permission.)

February 1934. The USDA started early. Gonzalo Merino of the USDA in the Philippines visited HSPA for advice on using parasitic wasps and flies to control the armyworm *Cirphia unipuncta* in sugar cane. Merino was impressed not only with wasps and flies but also toads. With Fred Denison's help he collected 27 toads from rice fields around Waipio, took them back to the Philippines[14] and released them near Manila. Fifteen years later, cane toads reached the southern-most island, Mindanao.[15]

---

14 Hawaiian Sugar Planters' Association 1934a.
15 Lever 2001, p. 40.

January 1935. HSPA sent toads to Formosa (now Taiwan) but none survived.[16] Two months later, Mungomery from BSES in Queensland arrived in Honolulu, sent by Bell to collect toads. This story is told in the following chapters.

Around the same time, GL Windred, Australian entomologist with the sugar company CSR in Fiji, was in Honolulu to collect parasites of a moth that attacked banana plants.[17] On his return to Fiji he recommended toads be imported. A year later, HSPA toads arrived at Lautoka on Viti Levu. From there, toads were distributed around Fijian islands and Funafuti atoll in the Ellis Islands. Four years later, toads were sent from Fiji to the Solomon Islands.[18]

In December 1936, Cyril Pemberton closed a circle of biogeography. In 1532, Martim Afonso de Sousa introduced sugar cane to the home of *Bufo marinus*; little more than 400 years later, Pemberton introduced giant toads to the ancestral home of sugar cane – the Pacific island of New Guinea. The main purpose of the trip was 'to get more cane material that would be useful in the Hawai'i breeding program' and Pemberton was joined by Colin Lennox, HSPA Assistant Genticist.[19] Pemberton had last collected seed from sugar cane in New Guinea in 1929, but this time there was a pleasant change. A year earlier this 49 year old bachelor had married Mildred Lucas, librarian at HSPA. On this expedition, Mildred accompanied Cyril, and Colin's wife Ginny accompanied him. No more lonely nights in shacks and grass huts in remote jungles. With wives in tow it was hotels, a wash and decent meals.

Rabaul, on the island of New Britain, was experiencing a plague of caterpillars from the sweet potato hawk moth *Hippotion celerio*. It was disastrous for indigenous people for whom sweet potato was a staple. Before setting out from Honolulu, Pemberton sent a shipment of toads from HSPA to the Department of Agriculture, Stock and Fisheries Experimental Station at Keravat about 25 miles (40 km) from Rabaul.[20] As insurance, he also arranged for toads to come from Queensland. BSES

---

16 Lever 2001, p. 47.
17 Hawaiian Sugar Planters' Association 1935b.
18 Lever 2001, p. 129.
19 Pemberton (n.d.).
20 Assistant Director BSES 1937b.

received a telegram from Pemberton: 'Arriving Cairns about February sixth en route to Rabaul. Can you give me thirty mature Toad then. Pemberton.'[21] Arthur Bell cabled a reply authorising the gift of toads.[22] Both lots of toads were released at Kerevat. From Rabaul on New Britain, Pemberton's toads were distributed around the country to clean up pests of native food gardens. Toads ended up thriving in many parts of New Guinea[23] but were particularly prevalent in coastal kunai grasslands that ring the island.

On the trip, Pemberton's wife Mildred became pregnant with their daughter Mary, and she discovered first-hand that the career of an entomologist was far more exciting than just microscopes and tweezers. The party decided to visit the islands of New Ireland and New Hanover to inspect indigenous cane plants, but a squall hit the small launch, it lost its sails, its motor would not start and the tiny craft was smashed onto a reef. The party made it to shore through the surf and lacerating coral and were eventually rescued from the remote atoll by missionaries.[24] Along with head hunters, crocodiles, tigers, cobras, malaria and innumerable bites and stings, Pemberton added shipwrecked to his list of adventures.

In May 1937, back home in Honolulu via Sydney, Pemberton wrote to Arthur Bell to thank him for the toads that were now 'thriving' in New Britain.[25] Thirty years later, when Australian herpetologist Mike Tyler visited Kerevat, it was a different tale. 'At night the emaciated toads ... wandered far out on the lawns and roads and ... were simply too weak to reach shade and so died in the early morning sun.'[26] Their muscles were wasted and Tyler supposed that the large population of toads had depleted insect food such that they were now starving.

In 1937, HSPA sent toads to Egypt. It was a tough assignment that no toads survived.[27] HSPA toads were also taken to Guam, west of the Hawaiian chain, to control scarab beetles. RG Oakley of the USDA in

---

21   Mungomery 1936.
22   Assistant Director BSES 1937a.
23   Easteal 1981.
24   Pemberton 1937a.
25   Assistant Director BSES 1937b.
26   Tyler 1976, p. 90.
27   Easteal 1981.

Guam collected 398 toads from Fred Denison. They proved popular. HSPA dispatched two more shipments[28] to Guam at the USDA's request.[29] A year later, HSPA entomologist Otto Swezey reported that slugs were much scarcer in Guam.[30]

The USDA also acted on its own. In Florida, closer to home, the USDA bypassed HSPA toads in favour of toads imported directly from Puerto Rico. In April 1936, they released 195 Puerto Rican *Bufo marinus* into the Florida Everglades to control wireworms. Ninety-five were released at the USDA's Belle Glade Station and 100 at their Canal Point Station. With native toads already present in Florida, *Bufo marinus* ran into predators familiar with toads, and their numbers increased only slowly. *Bufo marinus* did not trouble the wireworm population[31] but, instead, made their way into urban environments from Florida Keys to north of Tampa. Their numbers were boosted by additional *Bufo marinus* introduced from the Caribbean in 1944, 1957 and 1960.[32]

Toads were also introduced to cane fields in Louisiana in the 1930s, but none survived. If toads could have survived in Louisiana they would most likely have established themselves there naturally from southern Texas.[33]

In 1938, *Bufo marinus* from Puerto Rico were taken to the island of Mauritius[34] in the Indian Ocean, but were twice refused entry.[35] But around 40 years later the Indian Ocean was conquered when *Bufo marinus* took up residence on Diego Garcia, the southernmost atoll of the Chagos Archipelago.[36]

And war played its part in anuran biogeography. In the Second World War, wherever US Army troops went in the Pacific, toads went along, not discriminating friend from foe. They hid in equipment, hitched rides on transports, joined beachhead assaults, and occupied

---

28 Pemberton 1937b.
29 Hawaiian Sugar Planters' Association 1937.
30 Jepson & Moutia 1938, p. 383.
31 Ingram et al. 1938.
32 Easteal 1981.
33 Lever 2001, p. 66.
34 Jepson & Moutia 1938, p. 379.
35 Lever 2001, p. 49.
36 Lever 2001, p. 49.

new territories, Japanese or Allied didn't matter. Shell holes were perfect for breeding, blasted vegetation brought abundant insects close to ground and men and creatures killed in the mayhem soon hosted insect larvae. For toads it was just a matter of sitting and waiting for dinner. War was good to *Bufo marinus* and, because these toads appeared on islands during wartime, they are sometimes known as the 'American frog'.[37]

In 1949, the US Army took toads from Saipan to Chichijima in the Ogasawara Islands of southern Japan to control centipedes. Toads liked Japan and bred happily. From there, toads were introduced to other Japanese islands to control pests.[38]

Although Pemberton was the architect of the toad fantasy, many years later he reflected on the usefulness of the toad. He concluded, 'It is doubtful if it [*Bufo marinus*] contributed much in anomala [beetle] control'[39] but it did greatly reduce centipede populations in Honolulu, for which townsfolk were grateful.[40] Some residents were revolted by the creature but, although the occasional dog died, there appear to be no records of other animals on Hawai'i suffering as a consequence – apart from insects.

But it was very different when the toad fantasy spread to Australia.

---

37   Lever 2001, p. 118.
38   Lever 2001, p. 34.
39   Pemberton 1964, p. 709.
40   Pemberton (n.d.).

10
Toads for Queensland

Tip Byrne, Queenslander of Irish stock, remembered cane beetles of his youth:

> You'd be going into a paddock, picking beetles and putting them into a billy [can] and the cane leaves would be weighted right down with beetles on them everywhere ... millions, millions ... misery at the highest. People were out early in the morning ... walking through the cane paddocks, trying to pick these beetles. The beetles fly into the paddock at five o'clock and they're on the ground by half past five so you haven't got much time at all to do it ... they eat all the root systems out ... the cane cutters tried to cut it and they couldn't. There was no sugar in it. We had a prehistoric idea ... you put a knapsack on your back and you had a spear gun and put the spear into the ground and pump this poison into them and the poison was terrible ... you couldn't breathe it in because it would kill you. You'd walk into cane that was six foot [two metres] high and start pumping this stuff into the [cane] stools and you'd get lost ... back breaking and stupid in the first degree and never solved anything.[1]

---

1 Byrne 1988.

Chemicals, both noxious and obnoxious, had to be applied to the soil to kill the buried cane grubs. They included carbon bisulphide, paradichlorobenzene, white arsenic (arsenic trioxide) and Paris Green (copper acetoarsenate).[2] When cane grubs pupated and adult beetles emerged from the soil unharmed from the chemical onslaught, they rested on trees in rainforest remnants surrounding cane fields. So cane growers dusted these trees with arsenic compounds as well. But while these toxic chemicals laid waste to other creatures within close range, they were ineffective in controlling populations of either cane grubs or cane beetles. The non-chemical alternative, attempting to collect all grubs or beetles by hand, was expensive and also ineffective.

1928, Bundaberg, Queensland. Reg Mungomery, Assistant Entomologist with BSES, did his sums and reckoned that rather than picking grubs by hand from the soil, farmers should do nothing! Farmers should stop the back-breaking task of picking up cane grubs. Locals who had been banging their heads against a wall stopped, and they felt much better. They thanked Mungomery for his good advice. Both the local Cane Growers' Association and the Isis Shire Council[3] wrote letters of appreciation to the Director of BSES. Although there was still no effective control of cane grubs, locals were grateful to Mungomery for stopping the pain.

Mungomery was a local, born and raised among the cane farms of Childers, south of Bundaberg in central Queensland. His father was a blacksmith. He was a clever, sporty youth who played tennis, and the violin. At school in Childers in 1915 he excelled in biology and later in metallurgy at the Charters Towers School of Mines not far west of his home.[4] After finishing school he worked first on local cane farms then, in 1923, went to work for the Prickly-pear Board in Rockhampton, hired as an Assistant Entomologist by Alan Dodd – former BSES entomologist himself. Mungomery soon left the Prickly-pear Board, seduced by a job offer in the Mount Isa silver and lead works until the stark reality of the smelters changed his notions. He left the smelters after a year to join BSES. Schoolboy enthusiasm persisted into adulthood. He seemed perpetually enthralled, with an excitable stammer, an

---

2   Mungomery 1950, pp. 38–48.
3   Shire clerk, Isis Shire Council 1928.
4   Leverington 2006.

upright sheaf of wiry hair, a tall forehead and permanently arched eyebrows like a terrier waiting for a ball to be thrown.

By 1933, Mungomery was married, a house owner and making his name as an Assistant Entomologist in BSES. He was based at Bundaberg and reported to Arthur Bell in Brisbane, who was in charge of both plant pathology and entomology in BSES, and responsible for the war on cane grubs.

Eighteen months had passed since political favours had been traded to get Arthur Bell to Puerto Rico in the midst of the Depression. Now with new rumblings of war in Europe and a 'fifth column' of cane beetles already underground in the cane fields, pressure was on BSES and on Bell himself to perform for the sugar industry and for the economy of Queensland.

Monday 25 September 1933 brought early spring rain to Queensland – it had rained all weekend. Arthur Bell left his wife and his mother in the small wooden cottage in the leafy Brisbane suburb of Taringa and followed the rainwater running down the hill to the railway station. In the city, BSES offices were housed in the Department of Agriculture and Stock, on William Street between Parliament and the Treasury. That morning the offices were all damp wool and tobacco. Tweed-suited, brylcreemed gents smoked pipes and chippered the 'serviceable rains'[5] of the weekend. Good falls had been recorded right along Queensland's eastern seaboard and officers in the Department were upbeat about the prospects for beef, dairying, wheat, barley, potatoes and, of course, sugar. But early rains would herald the early return of cane beetles and their grubs. In Queensland's cane fields, pupating grubs would now complete their life cycles, dig their way out of warm moist soil and fly to neighbouring trees to mate before returning to the soil to lay eggs for a new brood. A new wave of grubs would hatch and join other cane grubs, some in their second season underground, eating roots of the new season's sugar cane.

At his desk, Bell followed sequences, cause then effect, rains then grubs, elation then dismay, BSES heroes then villains. Bell was back in the firing line and his staff had nothing to offer. The answer dawned slowly. There were two volumes on his desk. The first, HSPA's Annual

---

5   *The Courier Mail*, 26 September 1933, p. 14.

Figure 10.1 Arthur Bell, BSES. (BSES Limited. Used with permission.)

Report, told of *Bufo marinus*, a new control for white grubs. The second was the massive tome of proceedings from the Puerto Rico Congress. Recollections of feasts, dusty motor tours, hot days in cane fields, jacket off, talking diseases, pests, production, struggling with language. The little research station at Isabela, George Wolcott excited, animated, showing off *Bufo marinus,* giant toads, Raquel Dexter at his side. Her paper, somewhere in the leather-bound volume on the desk. And Cyril Pemberton took *Bufo marinus* to Hawai'i, successful by all accounts. Bell's conclusion: BSES really must do more in biological control, emulate HSPA and maybe try *Bufo marinus* against cane beetles in Queensland.

Bell penned a memorandum asking his entomology staff if they thought *Bufo marinus* would help control cane grubs.[6] Reg Mungomery's response was quick – he knew cane grubs first hand. He did not think toads would be much use. Mungomery elaborated for his boss:

> the habits of the cane beetles are such that this insect is not likely to be controlled by a predator on the adult stage, for the reason that adult females are in evidence for only ½ – ¾ hour on the night of their emergence, hence a predator would have to be particularly active and be present in very large numbers to effect any appreciable control.[7]

Moreover, the common green frog was already known to eat beetles with little overall impact on populations 'and before any attempt is made to introduce B. marinus, I think the habits of our common frog might be more fully investigated'.[8]

There was no hurrah for the toad from Mungomery.

Queensland's summer of 1933–34 was wet: good for rats, borers and cane grubs. The 1934 cane harvest was poor[9] with sugar production down 25% compared to the previous year.[10] BSES Director, Bill Kerr, complained that because his staff had nothing new to offer, 'the usual band of opportunists'[11] had taken advantage of the situation and

---

6   Pathologist BSES 1933.
7   Mungomery 1933.
8   Mungomery 1933.
9   Kerr 1934, p. 5.
10 Kerr 1935b, p. 5.
11 Kerr 1934, p. 62.

offered farmers supposedly innovative cures, like spreading common salt on grub infested fields. This was a cheap remedy made all the more attractive because BSES had not proposed it; an innovation for farmers – a money spinner for charlatans.

In March 1934, as a show of support for cane growers, Mungomery was transferred from Bundaberg to Meringa – 'the BSES grub station' – near Gordonvale, in the grub infested northern cane fields.

December 1934. After the poor cane harvest had been milled and with the moans of cane farmers in his ears, Arthur Bell's thoughts returned to biological control, to *Bufo marinus* and to Hawai'i. He discussed his ideas with BSES Director, Bill Kerr. He recounted the legendary success of HSPA in biological control and proposed that if Reg Mungomery were sent to Hawai'i perhaps some of this lineage and expertise might rub off on him. And perhaps *Bufo marinus* would perform as well in Queensland as it reportedly had in Hawai'i. Kerr proposed to the Minister of Agriculture and Stock, via the Under Secretary, that Mungomery be sent to Hawai'i to 'study the habits of the giant toad, *Bufo marinus*, the collection of suitable specimens and their introduction to Queensland for the purpose of control of the greyback cane beetle'.[12] Mungomery was also to enquire about tachinid flies, control of anomala beetles, control of rats, and 'make a general survey of biological control (which has been carried out so successfully in Hawaii) with a view to its possible extension in this country'.[13]

There was another reason for Bell and Kerr to send Mungomery to Hawai'i. In just eight months' time, in August 1935, BSES was to host the Fifth ISSCT Congress in Brisbane, the successor to the Fourth Congress in Puerto Rico, three years earlier. BSES wanted to show the rest of the sugar-growing world that it was at the cutting edge of science. The toad was the latest technology in biological control, its reputation was proven internationally, and there was a hope – if not certainty – that it would control cane grubs.

Reg Mungomery was the man for biological control. In his short time with the Prickly-pear Board he had been schooled in their philosophy and the 'safeguards' in their program. In 1934 Mungomery told Queensland cane growers that:

---

12   Director BSES 1934.
13   Director BSES 1934.

Figure 10.2 Bill Kerr, BSES. (BSES Limited. Used with permission.)

biological control does not consist in rushing off to a foreign country, bringing back a number of parasites, and letting them loose upon the unsuspecting pest ... such a project is not to be embarked upon lightheartedly, but only after the most mature consideration, since a false step may have most disastrous economic consequences through the upsetting of the whole biological balance.[14]

But despite his earlier misgivings about the effectiveness of *Bufo marinus* on cane grubs, Mungomery quickly became a toad convert. He read Raquel Dexter's paper from the Puerto Rico Congress, together with reports of the toad in the West Indies.

On 10 January 1935, Reg Mungomery wrote to Arthur Bell, reinforcing the value in sending him to Hawai'i. He told Bell that the toad would be beneficial in eating not only adult beetles, but also weevils, borers, caterpillars, and rats.[15] The promise of a trip to Hawai'i was the catalyst for Mungomery's conversion. Soon, there would be no stronger proselyte of the toad in Queensland.

On 15 March 1935, the Queensland Cabinet approved Mungomery's trip to Hawai'i with a budget of £200.[16] The next sailing of the Matson Line was just two weeks away. Quarantine matters were quickly arranged and Dr John Cumpston, Commonwealth Director General of Health in charge of quarantine, approved importation of toads.[17] It was unremarkable. At the time, Australia's Quarantine Act (1908) was 'a comprehensive set of national laws governing the control of infected persons, vessels, goods, animals and plants entering the country from overseas'.[18] It was designed to prevent diseases entering Australia rather than to prevent importation of new plants and animals that were essential to the development of Australian agriculture. Furthermore, BSES was a respected state government research institution. For Cumpston, importation of toads by BSES did not present a quarantine risk.

On Wednesday 3 April 1935, the SS *Monterey* of the Matson Line departed Sydney heads – 'a sovereign born to the homage of dipping palms ... a swift modern galleon and an ocean home of courtly elegance'.[19] Mungomery was sequestered in cabin class, akin to business class on an aircraft today. Over the 12 day voyage to Honolulu, this excitable young Queenslander rubbed shoulders with hopeful young starlets on their way to Hollywood, businessmen heading for deals in

---

14 Mungomery 1934b.
15 Mungomery 1935b.
16 Director BSES 1935e.
17 Director BSES 1935a.
18 Nairn et al. 1996, p. 288.
19 *The Brisbane Courier*, 3 May 1932, p. 4.

US dollars, and tourists on a cruise of a lifetime to New Zealand, Fiji, Samoa and Hawai'i.

On 28 April 1935, a Sunday afternoon, Mungomery was seated at a writing table in the Pleasanton Hotel, Honolulu – Cyril Pemberton's home when he was in town. Four things were on Mungomery's mind as he penned a letter to Arthur Bell. Firstly, he reported the death of a two year old child of Philippino workers on the outskirts of Honolulu. A guest had brought the family a large toad for dinner, one child had died and the other was in hospital, gravely ill. It was a giant American toad. Mungomery was worried about the reputation of the toads he was about to import. He suggested to Bell, that 'it would be a good plan to say nothing of this until I get the toads to Australia and then we can sound a note of warning through the press and the "Quarterly Bulletin" that they are not the edible variety of frogs, but that these toads possess certain poison sacks which render them extremely dangerous if they are eaten'.[20] This seems to be the first time that Mungomery understood that the toads were poisonous. But he did not understand the nature of their toxicity, did not consider the implications of releasing toxic toads into the Australian environment. He did not recall his own advice to cane growers about 'disastrous economic consequences' and 'upsetting of the whole biological balance'. He did not propose a change of plans. He just asked Bell to keep a very dangerous secret.

The second thing on his mind was an attack of appendicitis which had worried him, so far from home. But Pemberton had come to the rescue and taken him to his own doctor. Mungomery asked Bell, 'if you happen to see my wife in Bundaberg, please do not mention it to her as I do not wish to worry her unnecessarily. However, if you get word that I have to undergo an operation you will know the cause'.[21] Mungomery did indeed undergo an operation and spent 11 days in hospital in Honolulu, ministered to by Pemberton.[22]

The third matter was the arrangements for hosting the Fifth Congress of the ISSCT in Brisbane. Mungomery's gloomy news was that, because of the prevailing economic conditions, most of those from Hawai'i who were sending papers would be unable to attend in person.

---

20   Mungomery 1935c.
21   Mungomery 1935c.
22   Pemberton 1935c

It threatened the success of the ISSCT Congress in August and the international standing of BSES that they hoped to improve.

The fourth matter was a breeding pen for the toads at Meringa. In Hawai'i, Pemberton had not been able to breed toads in captivity in the laboratory. Instead, Denison had raised toads in ponds from strings of eggs collected from the rice field. But Mungomery, determined to breed toads, had drawn up plans for the pen and given them to his colleague at Meringa, James Buzacott. Mungomery requested Bell's permission to spend the money – Bell instructed Buzacott to start construction immediately.

Buzacott followed Mungomery's design for a pool inside a wooden framed octagon, sheathed and roofed with chicken mesh to keep potential predators of the toads at bay; there was no knowing what creature would like to eat a novel and precious toad.[23] A little way away from the buildings at the Meringa Experiment Station, he made a shallow excavation around three metres in diameter, lined it with concrete, strategically placed rocks around the edge, placed a sprinkler in the centre and planted hyacinth and water lilies. Around the edges he planted ferns and taro to provide shelter and shade and to make the toads feel at home. It was a palace, a Toad Hall, with all the comforts to induce toads to breed. The welcome mat was out for toads at Meringa.

Sunset, Saturday 1 June 1935, Honolulu. Toad snatchers were out. The streets not safe for innocent amphibians. The day cooling to evening, a family on their lanai, a ukulele, and the low trilling of giant American male toads. A car cruised slowly by and pulled up. On the roadway, street lamps cast pools of light where toads sat, waiting for food to drop in. Every few minutes a moth or beetle would fall, singed by the lamp, exhausted from flying in circles. When food landed, the nearest toad made a step or two, leaned forward, grabbed it and swallowed, helped along by its right hand. Three men got out of the car and crawled on hands and knees across the suburban lawn, dragging a cardboard box. They moved quickly, scampering on all fours, lunging, snatching toads and dropping them in the box. Reg Mungomery, Cyril Pemberton and Fred Denison snatching toads for Queensland.[24] Later, they snatched more toads from Denison's breeding pens at Waipio.

---

23   Mungomery 1935a.
24   Mungomery 1935a.

They snatched a total of 102 toads in equal numbers of males and females – not an easy task. Mungomery placed the toads in suitcases in the same way Pemberton had carried the toads from Puerto Rico – slots for air and moistened wood shavings for comfort.

Monday 3 June 1935. Mungomery and toxic companions left Honolulu housed in cabin class luxury aboard the *Mariposa*, sister ship to the *Monterey*, bound for Australia.

Monday 17 June 1935. The *Mariposa* docked in Sydney Harbour. *Bufo marinus*, giant American toads, toxic amphibians, arrived on the island continent. All documents were in order, Cumpston's permit was presented, and giant American toads freely entered Australia.

There was no official welcome. No brass bands playing. No demonstrating protesters. But there was still time to reconsider, to refuse entry. The toads were still in suitcases and it would have been easy to gas them – a peaceful death. But the executioner would have had to deal with Reg Mungomery, now the most passionate and dedicated guardian of precious toads.

Only one toad had died on the voyage and Mungomery respectfully disposed of the body. He fed and watered the survivors, explained to them their arrival on land, where they were headed, north on the train to Meringa, a long trip and a lovely home to go to. He changed trains at the Queensland border, fed and watered toads. At Bundaberg he collected wife Martha, cockroaches and water for toads. Husband and wife relaxed in the pullman car, Reg explained livid scar on abdomen, genuine Hawaiian grass skirt not appreciated by tearful Martha, so Reg checked amphibians, wrangled cockroaches, avoided awkward questions. Ten days of loving care later, and toads splashed happily in their pool at Meringa.

Monday morning, 1 July 1935. Mungomery, with fatherly pride, reported to Arthur Bell that at the weekend he had found 'a nice long string of toads' eggs in the water'.[25] Mungomery had thoughtfully given the toxic immigrants a delicious fly-blown bullock's head that generated abundant maggots for the toads.[26] They liked maggots, liked Queensland, liked Buzacott's handiwork and bred with no inhibitions. It was more than they had done for Pemberton! It was a triumph for

---

25 Mungomery 1935d.
26 Mungomery 1935d.

Figure 10.3 SS Mariposa departing Sydney Cove March 1932. (Australian National Maritime Museum, Samuel J Hood Studio Collection. Object no. 00034636. Used with permission.)

Mungomery and he adored them for it; ordered more bullock's heads, more treats for his toads.

Mungomery held true to his word about warnings not to eat toads and in the July 1935 edition of the *Cane Growers' Quarterly Bulletin* he wrote, with key words bolded: 'This toad, though large, is not the edible species of frog, **and it must not be eaten.**'[27] But the toad's toxicity was not publicised in local newspapers where the rest of the world could read about it, and snakes, goannas, quolls, kookaburras, cats and dogs could not read anyway. Mungomery's warning about the arrival of a toxic amphibian was a sham.

Sunday evening, 18 August 1935. After tennis, Reg Mungomery noticed the wind change – the chilly westerly had stopped and it was now blowing gently from the east, from the coast. Sunset gilded clouds building in the east; winter showers were on the way. Toads had made

---

27   Mungomery 1935e, p. 24.

full use of the pool and amenities in the love pad at Meringa, thousands of eggs had strewn the water and vegetation and toadlets were hopping all around Toad Hall. With rain on the way it was time to release them. The next day he would ride out from Meringa into the Queensland cane fields generously distributing toads for cane growers, far and wide.

Monday morning, 19 August 1935. Mungomery and James Buzacott counted toadlets into six sugar bags and tied them to the research station's bicycle. Mungomery dropped a hessian water bag on the crossbar, slung his knapsack across his back and pedalled off down the road. An amateur violinist, he conducted Schumann's *Traumerei* as he rode. In full voice, singing for his toads, shoulders back, holding the visceral climax aloft in cupped hand. A happy man – content with his cargo of toxic amphibians.[28]

Not far down the road, Mungomery came to Little Mulgrave River. The river was perfect, just three to four metres wide with a shallow gravelly bottom. It flowed east out of soft hills, lined with a narrow strip of rainforest left after the flood plain was cleared for cane. A constant chatter of birds, temperature and humidity rising, cicadas drummed as one. A flash of brilliant blue as an azure kingfisher flew straight and fast along the water, feeding on the wing. A pair of kookaburras, duelling hysteria, each out-laughing the other. Mungomery uneasy about the birds that could eat his toadlets. And snakes could feast as well. He tramped the water's edge to scare them off, took a sack of 500 toadlets and squatted among the sedge. He undid the twine, encouraged toadlets out of the sack, chatted to them, proud father. Toadlets propped on their little front legs, blinked in the sunlight, rehearsed life's choices. They fanned out, instinctively knowing the invasion drill. Mungomery watched. A long look around the creek for threats. Birds seemed to ignore them. Satisfied, Mungomery re-mounted, and pedalled off to find another creek.

It was a grand day to be distributing toads.

Mungomery's apocalyptic ride defiled Australia's 120 million years of isolation from other amphibians. All safeguards of the Prickly-pear Board had been forgotten in the rush to glory, the rush to release the 'proven' toad. No prior testing, no thought of consequences, no going

---

28   I do not know if Mungomery rode a bicycle that day; it is plausible and somehow appropriate. He did play the violin.

back. Toads were set on a collision course with Australia's post-Gondwanan fauna. It was executed with schoolboy enthusiasm, fortified with ignorance of toads' toxicity and with no thought for the consequences. Mungomery's sole mission was to serve cane growers and keep them happy.

It was Mungomery's folly. The act of an evangelist, not a scientist.

Mungomery distributed 2,400 toadlets around Gordonvale in northern Queensland.[29] That afternoon it rained and showers continued through the week. It was good for vulnerable toadlets in their new home and stopped Mungomery from fretting.

On Tuesday 27 August 1935, the Fifth ISSCT Congress hosted by BSES met in Brisbane. Cyril Pemberton represented the HSPA and reported on introduction of the toad to Oahu to control anomala beetles.[30] And, in George Wolcott's absence, Cyril Pemberton presented two of his papers[31] on white grubs and toads in Puerto Rico. The organising committee inserted a late paper from Reg Mungomery[32] on breeding *Bufo marinus*. That was the extent of technical papers on toads. But it was a wide ranging congress: delegates were there to talk about diseases of cane, taxonomy of insects, advances in sugar milling and to network and meet old friends. And there, in the front row of the official photograph of more than 90 delegates, taken on the steps of the Brisbane Town Hall, were old friends Arthur Bell and Cyril Pemberton. Bell, roundly full-faced, relaxed and smiling, wearing a daringly modern striped suit in contrast to all the plain suits on display. Pemberton meticulously turned-out in a three-piece grey suit, arms clasped behind him, a hollow cheeked ascetic with round-framed glasses and sporting a close-cropped widow's peak of receding hair.

Local dignitaries shared the spotlight of the Congress, talked-up Queensland in general and the sugar industry in particular. The Premier of Queensland, Scottish-born painter and decorator, William Forgan Smith,[33] Labor member for Mackay and friend of the sugar industry, unveiled a cairn at Ormiston south of Brisbane. It commem-

---

29 Release notes 1935-1950..
30 Pemberton 1935b.
31 Wolcott 1935; Tucker & Wolcott 1935.
32 Mungomery 1935a.
33 Costar 2006.

Figure 10.4 Reg Mungomery, BSES. (BSES Limited. Used with permission.)

orated the first commercial manufacture of sugar in Queensland, 70 years earlier, by fellow Scot, Captain Louis Hope. Forgan Smith boasted to delegates of the 'technologists' contribution to the welfare of the [sugar] industry', of Queensland's 'big stake in sugar' and how 'we are the only country in the world handling tropical production with such success'.[34] And Sir Raphael Cilento, Director General of Health and Medical Services in Queensland, proudly told delegates of 'an experiment of world-wide importance' that was happening in Queensland: 'that is to say, production of sugar by white labour – a unique phenomenon'.[35] He reported 'authenticated figures' which demonstrated 'that white men can live and thrive in the tropical parts of Australia, and that white women can accompany them without any loss of fertility, mentality, or physique'.[36] It was Cilento's touching endorsement of the White Australia policy.

And there were pressing international issues; foremost was 'The continuance of the acute economic depression which has enveloped the entire world and which has afflicted no section of industry more seriously than the great sugar cane industry.'[37] Evidence of this was the many delegates who could not attend because of economic conditions, and despondency about world-wide over-production in the sugar industry which the General Manager of CSR confidently put down to a world-wide lack of consumption.[38]

Post-congress, delegates boarded a train for Queensland's northern cane fields and sugar mills. Their field trip would finish with a triumphal flourish at the BSES research station at Meringa where Minister for Agriculture and Stock, Frank Bulcock, opened renovated research facilities and revealed the latest technology in biological control – the giant American toad. In full oratorical flight, praising BSES in front of the visiting international delegates, minster Bulcock 'made a trenchant attack on those organisations which, fortified with a colossal ignorance, were prepared to come forward and criticise the sugar industry.'[39] With toads looking on, it was a stirring call to arms by the Minister who held the purse strings for the organisation. But just a little over a week before, Mungomery, himself fortified with a colossal ignorance of the impacts of toads on native fauna and habitats, had released toads into the Australian environment.

---

34   Anon. 1935.
35   Cilento 1935.
36   Cilento 1935, p. 32.
37   Anon. 1935.
38   Goldfinch 1935, p. 27.
39   Anon. 1935.

There was no going back. BSES basked in the spotlight, ignorant of the havoc they had just created.

The Congress had gone well, BSES had acquitted itself on the international stage, and toads were in the cane fields. It was an achievement for the small team. But news of Mungomery's folly had spread. Not everyone was as happy about the giant American toad as Queensland's sugar scientists and cane growers. The bubble of post-congress euphoria was about to burst.

11
War on Canberra

Friday 8 November 1935, Brisbane, subtropical Queensland, a humid morning. At 95 William Street, home of the Department of Agriculture and Stock and BSES, a telegram boy leant his bike against the railings, removed his cap, tugged his jacket to order and entered the porch. Bill Kerr, Director of the Bureau of Sugar Experiment Stations (BSES), opened the telegram and sighed. Dr John Cumpston, in charge of Australia's quarantine, had banned further release of giant American toads. Retired New South Wales government entomologist, Walter Froggatt, had complained. He convinced Cumpston that 'this great toad, immune from enemies, omnivorous in its habits, and breeding all the year round, may become as great a pest as the rabbit or [prickly-pear] cactus'.[1] Froggatt was determined to stop the toad.

The telegram was a shock, but not unexpected.

Cyril Pemberton, Hawaiian Sugar Planters' Association (HSPA) entomologist, had provoked his old friend Froggatt. Their close friendship went back 15 years to when Pemberton, newly recruited to HSPA, visited New South Wales to collect fig wasps and parasites of fern weevils. Pemberton last visited Froggatt in Sydney in October 1935[2] at the conclusion of the ISSCT Congress hosted by BSES in Brisbane. Pemberton was on his way home to Hawai'i when he told Froggatt

---

1 Froggatt 1936b, p. 9.
2 Pemberton 1935c.

Figure 11.1 Walter Froggatt at left, captioned 'Scientific staff of the Sugar Planters Association Honolulua. Chambers, Kotinski, Swezy, Davis' c. 1908. (State Library of New South Wales. Call no. PXB 393/1. Used with permission.)

of the toads breeding at Meringa. Froggatt was horrified. Pemberton warned Mungomery at BSES that 'Froggatt … is of the opinion that no comparison can be made between its [the toad's] future in Australia as against Hawaii. I can't change him. He is too mature a man for that.' Pemberton added for Mungomery's benefit, 'You have no reason to worry over this anti-Bufo sentiment. I count Mr. Froggatt one of my best friends in Australia and respect his attitude even though he is on the wrong track.'[3]

Walter Wilson Froggatt was no lightweight. He was founder and president for a record 11 years of the Naturalists' Society of New South Wales. He was a member of the Australian National Research Council, Fellow of the Linnean Society of New South Wales, the Royal Australian Historical Society, and the Royal Zoological Society of New South Wales. He was also founder of the Australian Wattle League, the Gould League of Bird Lovers of New South Wales, and the Wildlife Preservation Society of Australia.[4]

---

3   Pemberton 1935a.
4   McDonald 2006

11 War on Canberra

Figure 11.2 Walter Froggatt, Linnean Society of New South Wales, 1923. (State Library of New South Wales. Call no. PXE 884/7. Used with permission.)

In 1907–08, Froggatt travelled to Hawai'i, Barbados and Mexico investigating insect pests including the fruit fly. On that trip he spent time in the field with HSPA entomologists. As Chief Entomologist of New South Wales he had been responsible for knocking crazy proposals for biological control on the head. Proposals like importing red meat ants from South Africa to destroy baby rabbits in their burrows; for ferrets, stoats, weasels and mongoose to exterminate rabbits; and for vultures from Texas to control blowflies – by picking carcasses clean.⁵ Giant American toads were in the same league of crazy ideas.

Now retired, Froggatt detailed his complaint in a letter to the *Sydney Morning Herald*, but 'the newspaper refused to publish it on the grounds that it was grossly alarming in its nature, and that there must be something of value in the toad since the Commonwealth Government had permitted its entry to Australia'.⁶ But Froggatt managed to get a warning printed in Melbourne's newspaper, *The Argus*.⁷

Froggatt's complaint reached Sir Albert Cherbury David Rivett, recently knighted and Chief Executive Officer of the Council for Scientific and Industrial Research (CSIR) – the forerunner of today's Commonwealth Scientific and Industrial Research Organisation (CSIRO), Australia's peak scientific research body. David Rivett contacted Robert Veitch, Chief Entomologist of Queensland, also a Fellow of the Linnean Society of New South Wales. And Froggatt enlisted the Government of New South Wales to tell Canberra to prohibit the toad.⁸ Froggatt's complaint eventually reached Cumpston in Canberra who had given permission to import the toad. Froggatt's warning and his high standing made Cumpston change his mind about the giant American toad, and ban further releases. Cumpston explained that although he had permitted the initial importation of the toads for experimental purposes, 'since then the Department [of Health] has become worried by the possibility that the toads might become an even greater pest than the cane beetles they were to destroy'.⁹ Cumpston's concern was prescient.

John Howard Lidgett Cumpston, tall, thin, bespectacled, acerbic, 55-year-old was the first Director General of the Commonwealth Department of Health and Director of Quarantine, appointed in 1912.¹⁰ He led treatment and prevention of tuberculosis, cancer, venereal diseases and infant and maternal care, and promoted Australia as 'a paradise of physical perfection'. He had made his decision about the toad

---

5   Froggatt 1936a.
6   Director BSES 1935c.
7   *The Argus* (Melbourne), 25 June 1935.
8   Froggatt 1936a.
9   *The Courier Mail*, 30 November 1935, p. 10.
10 Roe 2006.

11 War on Canberra

and told the Queensland Under Secretary of Agriculture and Stock he would not be moved.

Saturday 9 November 1935, William Street, Brisbane. Bill Kerr distilled the essence of Froggatt's complaint into a memorandum to brief the Queensland Under Secretary for Agriculture and Stock:

> Mr Froggatt predicts the extermination of ground nesting birds, many lizards, certain frogs which supply water for famished blacks in the interior, some of the rare and peculiar insect life of Australia, etc. etc. [sic] He anticipates a tremendous epidemic of toads, which will greatly alter the fauna complex over much of tropical and semi-tropical Australia.[11]

In Kerr's mind this was a preposterously apocalyptic vision and he assured the Under Secretary that 'in Puerto Rico and Hawaii, where these toads have been bred as extensively as is possible, nothing but good has resulted'.[12] Kerr's defence of the toad was a memorandum of hyperbole.

Monday 11 November 1935, Armistice Day, Brisbane. At 11am, city traffic stopped mid-street. Trams came to a standstill. Drivers and passengers emerged and stood next to their vehicles. At the William Street offices of the Department of Agriculture and Stock the clatter of typewriters ceased. No phones rang. With poppies in lapels or on blouses, office workers rose to stand heads bowed for two minute's silence – one minute remembering those who had died on the Western Front, the other minute remembering the living. For returned soldiers, memories flickered behind closed eyes. With thoughts of war Kerr called in reinforcements and went on the offensive. The pride of BSES was under threat.

He wrote to David Rivett, CEO of CSIR in Melbourne, explaining that Pemberton was supporting importation of the toad by BSES 'which was made with the permission of the Director-General of Health, Dr Cumpston'.[13] Rivett replied, 'I am very glad to know that Mr C E Pemberton is prepared to defend the wisdom of the importation despite the

---

11   Director BSES 1935c.
12   Director BSES 1935c.
13   Director BSES 1935g.

decidedly pessimistic forecast of the New South Wales entomologist'.[14] Kerr had found a battalion of support in CSIR.

David Rivett's support was self-serving – but that mattered little in Kerr's battle. CSIR was 'very keen to use *Bufo vulgaris* [now called *Bufo bufo*] in an effort to deal with *Oncopera* [caterpillars, moth larvae] and other grass grubs'.[15] Rivett was worried that a ban on the giant American toad might interfere with his own toad program. Self-interest within Australia's peak scientific research organisation gave BSES a powerful ally.

CSIR's proposed use of *Bufo bufo* is not as outlandish as first appears. In 1937, two young men in Euroa, Victoria, bought into a franchise – they purchased breeding pairs of the American bullfrog *Rana catesbeiana* from the American Frog Canning Company in New Orleans.[16] In return for breeding them, the company would buy back mature toads for processing into frogs legs and other edible products. Although the bullfrogs and their venture never reached maturity, there were no apparent quarantine restrictions to prevent importation to Australia of at least three batches of breeding frogs. *Bufo bufo*, *Bufo marinus*, *Rana catesbeiana* were treated as agricultural imports in much the same way as breeds of cattle.

Robert Veitch, Queensland's Chief Entomologist, a dour, hard-headed Scot,[17] rallied behind BSES. But that was to be expected of the man employed by the Queensland Government, loyal to his colleagues in the Department of Agriculture and Stock, friend of cane growers. Like Rivett, he supported Kerr and importation of giant American toads.

Kerr called his regulars to arms and announced the ban to his loyal troops in the Australian Sugar Producers' Association and Queensland Cane Growers' Council. Bill Doherty, Council Secretary, immediately cabled battle plans to his troops: 'Federal authorities prohibiting further release giant toads in sugar areas Suggest strong agitation against Federal authorities stop Understand fullest enquiries made regarding toads before being imported.'[18]

---

14   Rivett 1935.
15   Rivett 1935.
16   Hince 2011.
17   Gill & Lloyd 2006.

Across Queensland cane growers formed up in ranks behind BSES, voiced outrage at federal interference and fired shells of hyperbole from newspaper howitzers at Cumpston;[19] 'toads are the best economic contribution to the sugar industry for many years'[20] and 'much time and money have been spent in investigating the habits of the giant toads'.[21]

But Cumpston was firmly entrenched.

Meanwhile it rained on the battlefield. Twenty-five millimetres of rainfall (an inch) fell on cane fields around Little Mulgrave River, filling pools, ponds and dams and swelling creeks. Havoc ensued. Giant American toads began spawning. Each adult female could lay around 30,000 eggs in fresh pools of rain. Toads were on the loose and it was now impossible to get the tadpoles back in the jar. Kerr knew it and told Mungomery that the Department of Health 'will be certainly put to some trouble to round up and destroy those toads which have been liberated to date'.[22]

It was a curious situation. Kerr knew it would be impossible to capture toads that had been released, yet he worried that thousands of tadpoles in the pool at Meringa would have to be destroyed if the ban was not lifted.[23] Neither Kerr nor Mungomery understood that the toads already released would multiply rapidly, that they would colonise cane fields of their own accord and that tadpoles in Meringa were already redundant. But for Kerr and Mungomery, toads were precious and vulnerable, needing protection.

Mungomery had an idea. He agreed with 'the impossibility of trying to recover all the toads in the districts where we have liberated them' but thought it not unreasonable 'if they allowed us to make further releases in the same districts as where the other liberations were carried out'.[24] Of course, toads already liberated would multiply of their own accord, but Kerr, Bell and Mungomery had been slighted and they would not accept defeat from Froggatt. And Mungomery, wanting to

---

18 Doherty 1935.
19 *The Courier Mail*, 29 November 1935, p. 19.
20 *The Queenslander*, 28 November 1935, p. 19.
21 *Sydney Morning Herald*, 29 November 1935, p. 11.
22 Director BSES 1935d.
23 Kerr 1935a.
24 Mungomery 1934a.

see more toads on the ground in cane fields, could not think of destroying toads at Meringa. It became the new cause – allowing BSES to continue releasing toads where they had already been liberated in the areas around Cairns, Gordonvale and Innisfail. It was reasonable. The damage had already been done.

On Wednesday 20 November 1935 at headquarters, 95 William Street, Brisbane, Bill Kerr, advised by gunner Arthur Bell, readied his heavy artillery. He navigated wooden stairs, balconies and arched corridors for a council of war with Arthur Graham and Frank Bulcock – respectively Under Secretary and Minister for Agriculture and Stock. Headquarters agreed BSES had done no wrong and planned an artillery engagement to destroy Cumpston's position, to drive him to the rear. The first volley, registering the target, was a letter from Graham to Cumpston proposing that 'permission be granted for the release of young toads in those areas where liberations have already been made'.[25] Then came the heavy bombardment. The Director of BSES coached the Under Secretary in drafting a letter to the Minister, who wrote to the Queensland Premier, who wrote to the Prime Minister.[26] It was the same letter, almost verbatim, retyped on three letterheads. Bill Kerr's assurances were repeated along the chain of command. The Queensland Premier relayed Kerr's words to the Prime Minister:

> The entomological literature from the sugar world is unanimous in its opinion respecting the value of this animal, and not one valid criticism has been levelled at it. In Puerto Rico and Hawaii it is in great demand in both field and town.
>
> It is certain that the toads which have already been liberated could never be recovered, and the animals will continue to multiply. It is felt, therefore, that no objection could be raised against our proposal that permission be granted for the release of the toadlets which are produced at Meringa, in those areas in which liberations have already been made.[27]

---

25   Undersecretary Queensland Department of Agriculture and Stock 1935.
26   Secretary Queensland Department of Agriculture and Stock 1935.
27   Premier of Queensland 1935.

11 War on Canberra

Figure 11.3 The Hon. Frank William Bulcock, 1942. (State Library of Queensland. Negative number 104058.)

2 December 1935, Canberra. Federal Parliament was still sitting[28] and Prime Minister Joseph Lyons was in the House. The response was immediate. Lyons over-ruled Cumpston. Outgunned, Cumpston surrendered and the white flag went up. He cabled the Queensland Under Secretary on 4 December to 'Release further toads in areas where liberation already taken place in accordance terms your letters.'[29] The Prime Minister later confirmed his decision in a letter to the Queensland Premier.[30]

The campaign was won in less than a month. The guns fell silent and Cumpston retreated to dress his wounds, and to shed blame. He

---

28 *The Courier Mail*, 2 December 1935, p. 1.
29 Cumpston 1935b.
30 Prime Minister of Australia 1935.

Figure 11.4 Queensland Department of Agriculture and Stock, 95 William Street, Brisbane, backing onto Queen's wharf on the Brisbane River, June 1936. (Queensland State Archives. Digital Image ID 461.)

wrote to Under Secretary Graham: 'It is recognised, of course, that your Government is prepared to accept its share of responsibility for the action which is now being taken.'[31]

Neither the Prime Minister of Australia, nor the Queensland Government, and neither its Department of Agriculture and Stock nor its Bureau of Sugar Experiment Stations ever took responsibility for their actions. Cumpston's pointed comment to the Queensland Government became lost in the margins of history.

Support for the giant American toad went right to the top of government. BSES Director Kerr, Under Secretary Graham, Minister Bulcock, Premier Forgan Smith and Prime Minister Lyons championed the giant American toad.

In December 1935, there were celebrations at 95 William Street. With friends in high places, Christmas came early for BSES. Kerr congratulated his troops and the allies that helped him win the battle. He wrote to the Secretary of the Australian Sugar Producers' Association: '[I] feel sure that your representations have assisted materially in effecting this modification of the prohibition.'[32] And he wrote to David Rivett

---

31 Cumpston 1935a.
32 Director BSES 1935f.

of CSIR: 'In view of the substantial evidence which exists regarding the virtues of this animal, I feel sure that Dr Cumpston's investigations will reveal nothing of a calamitous nature ... I trust also, that you may be favoured, and that *Bufo vulgaris* will prove an effective control agent for *Oncopera*.'[33]

Walter Froggatt's complaint against the toad became public in January 1936 when his article was published in *The Australian Naturalist*.[34] He wrote:

> All our ground fauna will become their prey, and all our curious, mostly harmless, and often useful ground insects, in forest and field, will vanish. The eggs and nestlings of all our ground nesting birds will be snapped up by these night-hunting marauders. All our frogs and lizards, most valuable insectivorous creatures, will be in danger of their lives.
>
> There is no limit to their westward range, and though originally natives of tropical regions will probably adapt themselves to our mountain ranges, and even reach the river banks and swamp lands of the interior ... this great toad, immune from enemies, omnivorous in its habits, and breeding all the year round, may become as great a pest as the rabbit or [Prickly-pear] cactus.[35]

Froggatt was prescient – Mungomery was indignant. Mungomery ranted to Bell:

> It is rather pathetic to notice that he [Froggatt] is so overwhelmed by an incurable bias, and that the statements contained in his article are so crammed full of inaccuracies, as to render his criticisms quite worthless ... Is he fearful of the possibility of the Giant Toad proving of some real value to the sugar cane plantations? ... It must be rather disappointing to him to realise that his protest is somewhat belated, for already Bufo marinus is present in thousands in the North Queensland cane fields, and moreover, Bufo marinus is now here to stay.[36]

---

33   Director BSES 1935b.
34   Froggatt 1936b.
35   Froggatt 1936b, p. 9.

Indignation was contagious. Bell complained: 'It seems unreasonable that we are still ignorant of the charges which have been levelled at these animals.'[37] And he wrote to Froggatt requesting the authority for statements in his article 'which casts a very definite shadow upon the professional integrity and ability of our entomological staff'.[38] It was a silly letter – petty indignation from BSES with no real defence for the toad.

Froggatt, measured in response, replied to Bell in a cursive hand on the letterhead of the Naturalists' Society of New South Wales, graced with gum leaf emblem, noting 'President: W W Froggatt'. He invoked 'my friend Mr Pemberton who is an authority on this toad's activities in Honolulu. I, myself, as an economic zoologist, know also something of the structure, fecundity, habits and life history of toads'.[39] His passion constrained in the dimensions of the page and by protocols of the times, he continued: 'We consider that no organisation or even a single state of the Commonwealth has the right to independently introduce such a possible menace to the continent as *Bufo marinus*.'[40] His actions were a duty that 'all nature lovers and zoologists owe to the country we live in'.[41]

Froggatt's complaint was lost in the clamour of cane growers for toadlets and in the continued lobbying by BSES to have the ban lifted completely so toads could be released anywhere in Queensland – the objective of their importation.

Wednesday 8 July 1936, William Street, Brisbane. BSES Director Kerr helped Under Secretary Graham to load the final barrage in the war with Canberra. Five rounds went in the volley.[42] The first round came from Froggatt's former friend, Pemberton, who gave a glowing reference for the toad:

---

36   Assistant Entomologist BSES 1936.
37   Cumpston 1935a.
38   Assistant Director BSES 1936c.
39   Froggatt 1936a.
40   Froggatt 1936a.
41   Froggatt 1936a.
42   Undersecretary Queensland Department of Agriculture and Stock 1936.

> I am surprised that a man of Mr Froggatt's standing as a naturalist could entertain such radical and biologically impossible apprehensions respecting so beneficial and innocuous creature as <u>Bufo marinus</u>. It is further surprising to me that he should write in such a vein, since I discussed the whole matter with him at great length on several occasions. I told him of the objections raised in Honolulu against it by a few neurotic old women and an occasional chronic pessimist, who cared nothing for the benefits accruing to the general community, but only conjured in their own selfish minds dire results to their nervous systems should one of the "loathsome creatures" cross their path in the rose or vegetable garden when in pursuit of a cockroach, centipede, scorpion, or pestiferous beetle. [43]

Pemberton explained that despite being crushed on roads they were not swarming all over Honolulu. Toads were in demand and 'the effect on the gardens is highly beneficial ... vegetable gardeners and flower growers are especially grateful'.[44] He followed with a tutorial on toad biology, explaining that 'Toads do not live on vegetation, fish, the air, the rocks, lizards, snakes, or frogs [which] prevents this toad ... from multiplying beyond a definite point. ... A biological balance is inevitable.'[45] He concluded that 'To discuss the objections [of Froggatt] further becomes ridiculous.'[46]

This contemptuous dismissal of Froggatt's complaint by such a world-renowned entomologist as Pemberton was designed to target laymen and politicians alike – to give them unswerving confidence in their support of the toad.

The second artillery round of American manufacture was a further character reference for the toad from Dr FX Williams of HSPA who had studied the toad in Guatemala.

The third round, a detailed, petulant critique of Froggatt's article, was fired by gunner Bell.

The fourth round was a review by BSES staff of 'all known literature' on the toad.

---

43   Pemberton 1936.
44   Pemberton 1936.
45   Pemberton 1936.
46   Pemberton 1936.

The fifth and final fragmentation round came from gunner Bell once more, now Assistant Director of BSES. It was full of toad excrement! Toads were considered precious and staff at Meringa were reluctant to dissect them to see what they had been eating, so they collected their droppings instead and dug around in them. They declared that most of the insects in the droppings were 'harmful species'.[47] And for good measure, perpetuating Dexter's myth, Bell sent Cumpston another copy of her paper on the feeding habits of *Bufo marinus*.

Cumpston could not return fire. He lifted the ban completely in September 1936,[48] made no requests for additional studies and placed no further constraints on the release of toads. BSES continued to breed and distribute thousands of giant American toads around sugar growing areas of Queensland. Their efforts were supported by the Premier of Queensland and the Prime Minister of Australia.

Reg Mungomery's initial reaction in 1933 had been right. Toads did not control cane beetles or their grubs. The task was beyond them. In 1921, BSES's own entomologists, James Illingworth and Alan Dodd, had documented the impossible task that a predator, like the toad, would face in trying to consume millions of cane beetles in two brief half-hour periods before dawn and after dusk, and only on days in the mating season when the beetles were above ground. But the prospect of an exciting overseas trip to Hawai'i befuddled Mungomery's reasoning and it stayed befuddled despite the emergence of the ineffectiveness of the toad in controlling cane beetles.

Thanks to BSES, Froggatt's complaint went largely unheard. Yet it was prescient. Toads spread out of cane fields and his predictions became horribly real. Australian residents, animals and people alike, soon had to learn to live with the toad.

---

47   Assistant Director BSES 1936b.
48   Assistant Director BSES 1936a.

12
Living with Bufo

At first, giant toads were a curiosity.

In August 1937, Queenslanders could not get enough of toads. A pair of them was displayed at the annual Brisbane Exhibition – affectionately called the Ekka. In a crafted diorama of cane farms, a curved amphitheatre supported a mural of jungle-clad mountain ranges and misty blue peaks. In front of the mural, maintaining perspective, stood a model of a sugar mill with trucks on a painted roadway and a cane railway stretching away through textured cane fields. And in a pond in the centre squatted two giant American toads, larger than the sugar mill, larger than life, 'great ugly things'.[1] Within the decade, Brisbane locals would be able to see toads on their own doorsteps.

Toads migrated out of cane fields and towards food, shelter and habitation – where humans tried to kill them. This distressed Reg Mungomery, self-appointed toad custodian for BSES. Around Gordonvale, young boys were throwing stones at toads, whacking them along the road with sticks, and popping at them with lever-pump air guns. Mungomery wanted to call the police about one boy who 'kills any toad that comes near his place, his only reason for doing this being because he hates the sight of them'.[2] Mungomery thought this unreasonable. He wrote to Arthur Bell to see what legislation could be invoked to protect

---

1 *The Queenslander*, 19 August 1937.
2 Mungomery 1938.

his precious toads,[3] and Bell made inquiries to the Under Secretary for Agriculture and Stock.[4] But Queensland's legislation was sadly lacking. There was no legal protection for the toad.

For BSES, giant American toads were precious and needed protection but no one seemed to understand how poisonous they were. When *Bufo marinus* is mouthed by a predator, poison – bufotenine – is secreted from parotoid glands on the shoulders. Regardless of age or sex, from eggs to adults they are toxic in all stages of life. Their poison causes sudden death. One chillingly clinical report retells of how an 11 year old girl in a small village near Iquitos, Peru, watched her mother and sister die:

> Early in the morning her brother had gathered numerous strings of toads eggs … [and] boiled them in water, after which she, her brother, her mother and the 4 year old sister ate them in a soup … [after about 30 minutes] she complained of stomach pains and told her mother, who by then realised that they had eaten toad eggs and said they were all going to die. [They all vomited, and after about 90 minutes] the mother and sister were prostrate, and their lips and fingers were blue: pustules had formed on her sister's lips. The abdomens of both the mother and sister appeared bloated and their bodies were hot and quite rigid. They were pronounced dead by 12 noon.[5]

The young girl and her brother ate little of the soup and they recuperated within a day or so. The eggs were identified as those of *sapo buey*, otherwise known as *Bufo marinus*. Her mother realised the danger too late.

Mungomery witnessed a report of similar deaths in Hawai'i but asked Bell to keep it a secret. There were other early warnings. In Hawai'i in June 1935, there were confirmed reports that dogs had died after mouthing toads. But there was also evidence that toads ate the highly poisonous black widow spider, recently introduced to Hawai'i. About the dogs, Cyril Pemberton reasoned, 'Should the toad ultimately

---

3 Mungomery 1938.
4 Assistant Director BSES 1938.
5 Licht 1967.

12 Living with Bufo

play an important part in suppressing this [black widow] spider in the field, such benefit alone will amply atone for the death of a few dogs.'[6]

At the start, Australians were ignorant of the danger. When, in 1937, a 'pickled, large-size giant toad' was displayed in Sydney,[7] a newspaper warned readers that the toad was poisonous to eat, 'but its warty appearance is likely to prove sufficiently repugnant to protect it from being "potted" by the blacks or mistaken for the edible frog by whites'.[8] There was nothing else to worry about.

Predictably, dogs died in Queensland: stray dogs, working farm dogs that knew no better, and pet dogs, especially terriers that could not resist the chase and catch. In 1937, farmers south of Cairns complained about dog deaths to Johnstone Shire Council who complained to the Minister for Agriculture and Stock. Arthur Bell, briefed by telegrams from Reg Mungomery[9] on the deaths, was called in to explain to Under Secretary Graham. Bell gave a rosy account[10] and mocked reports that a greyhound had been attacked and poisoned by a toad. This information was fed to Minister Bulcock and used by him in defence of the toad.[11] BSES's support for the toad did not waver despite mounting protests. Bell remarked to Mungomery, 'There exists a section who would do all they can to blackguard our imported friend, but whatever happens, the toad is with us and will remain.'[12]

He may never have said a truer word.

Complaints became louder, accompanied by squawks. One morning, Bill Thomas, licensee of the Central Hotel in Gordonvale, heard a commotion. He forgot to put the ducks away. Out in the yard a duck was on the loose. It was running around with a toad clasped firmly in its beak and was being cursed by the yardman. Bill joined in, stirring the dust. The duck ran in front, twisting and turning this way and that, wings at half stretch, webbed feet paddling the yard, neck outstretched holding the precious trophy. Bill followed behind, unlaced

---

6 Hawaiian Sugar Planters' Association 1935a.
7 Director BSES 1937a.
8 *The Daily Telegraph* (Sydney), 1 June 1937.
9 Bell 1937; Mungomery 1937.
10 Director BSES 1937b.
11 *The Queenslander*, 26 January 1938.
12 Bell 1938.

boots flapping at ten to two, elbows bent for balance on the turns, shirt tail fluttering in the breeze. There was no finer impersonation of a duck.

But the chase stopped. The duck stood still. It dropped the toad and began to flutter, stretch its neck and make little jumps. Soon the duck was rolling on the ground thrashing its legs and wings and, in just a minute or two, it was dead. Bill went over to examine the toad. It was about as long as his index finger, brown and warty, and white stuff was oozing from pads over its shoulders. It was still alive. It was the ugliest creature he had ever set eyes on and he reckoned that the white stuff would not be fit for man or beast. He picked up a spade lying against the fence and, with one swipe, flattened the mongrel toad. Later, after settling the dust in his throat with a beer, Bill picked up the phone in the bar and asked Mavis at the telephone exchange to connect him to Reg up at 'the research'. That day, the duck's demise was reported to BSES head office in Brisbane.[13]

In 1940, five years after Mungomery's folly, Kerr wrote in the BSES Annual Report that toad populations were at 'saturation point' around the cane fields but 'the number of beetles destroyed by the toad is relatively small when compared with the numbers that are nightly on the wing'.[14] The Annual Report contained the first admission by BSES of failure of the toad in its assigned task. It was beginning to dawn on them that the giant American toad was not performing as expected, was not controlling cane beetles and was doing very little apart from being a nuisance. Toads did not go out of their way to eat cane beetles and Bell confided to Mungomery that 'the problem appears to be to bring the toad and the greyback [beetle] to close quarters'.[15]

Doggedly, Bill Kerr, Arthur Bell and Reg Mungomery continued their defence, believing the toad just needed longer to demonstrate its powers.

It was a similar story in Hawai'i where the toads failed to control anomala beetles.[16] And in Puerto Rico, white grubs returned as a problem for cane growers[17] even though the giant American toad had been

---

13   Mungomery 1939.
14   BSES 1940, p. 17.
15   Assistant Director BSES 1939.
16   Pemberton 1964, p. 709.
17   Wolcott 1948.

assisted, for a number of years, by the common American bullfrog *Rana catesbeiana*. The result of the country-scale experiment, replicated in Puerto Rico, Hawai'i and Queensland, was that the giant American toad would not control populations of cane grubs, or indeed entire populations of any particular insect.

1946 saw the worst plague of cane grubs in Queensland in 20 years. The toad had been no help and had become a pest itself. But 1946 heralded a new wonder chemical, an organochlorine insecticide called benzene hexachloride (BHC), sold under the name of Gammexane.[18] Jim Buzacott at Meringa had done the experimentation. It just had to be dusted onto soil and cane grubs would die. BHC quickly became the weapon of choice. Cane yields increased, farms and farmers recovered, cane grubs were vanquished – up until 1987 when organochlorine insecticides were withdrawn from use in Queensland after persistent and dangerous residues were found in cattle.

Cane grubs were controlled for a period, but toads became pests in their own right. They multiplied and moved out of Queensland's cane fields and eventually moved out of Queensland.

1948, Innisfail, north Queensland, late evening. A brand new cream coloured Holden FX drove slowly from side to side down the road; a big curvaceous sedan with a prominent nose, Pancho moustache chromed grill, and glowing headlights in rounded fenders. It weaved from one kerb to the other, like a drunk; the driver silhouetted through the upright split windscreen, broad brimmed hat, hunched over the wheel, concentrated on the road. Street lamps dropped puddles of light, left, right, left. The ground moved under each – thousands of cane toads.[19] The driver reduced speed to improve accuracy, moved to the right, lined up driver's side front wheel from 10 metres out, adjusted the line, accelerated from five metres then pop, pop, pop. He swerved to the left and the next lamp, better accuracy this time, improved the score, pop, pop, pop, pop.[20] A new sport for Queenslanders, toad popping, and full points if you hit the toad head on because it went off with a bang.

---

18 Griggs 2005.
19 *Sydney Morning Herald*, 2 August 1947.
20 Adapted from account from Brent Vincent, resident of Cairns, interviewed in Lewis 1988.

Complaints continued – of chickens dying because of toads in their pens, of toads camping around beehives catching bees as they landed.[21] This was not news to Cyril Pemberton – he knew that toads ate bees in Hawai'i[22] as well as in Antigua.[23] But there was no warning for Australian beekeepers who were shocked to find a toad with 500 bees in its stomach.[24] They petitioned the Commonwealth Government to prohibit toads entering New South Wales.[25] But it was too late – toads were on the move and could not be stopped from vacuuming insects in their path.

In 1935, Queensland's Minister for Agriculture and Stock had vigorously defended giant American toads to the Prime Minister of Australia but, 14 years later, Maryborough City Council, 250 kilometres (100 miles) north of Brisbane, was asking for the Minister's help in eradicating them.[26] The problem was the tsunami of toads at the invasion front as they moved out of the cane growing areas, each trying to out-hop the other – a sea of moving toads. Things settled down a little once the front moved through, but the initial wave was a shock to townsfolk. In 1949, toads arrived in Brisbane, the subtropical capital of Queensland: 'With every step that you take toads are roving in front and on each side of you: a slowly moving sea of dark, shuffling hopping creatures … on the verandah and on the stairs … every morning the roads are littered with the crushed bodies of toads.'[27]

Despite the urban invasion Mungomery still defended giant American toads; defended them in the face of complaints of dogs dying, of toads eating pets' food, squatting in their water bowls, of people frightened by ugly warty toads when they stepped outside. He told listeners to a Brisbane radio station in March 1949 that the toad 'is pulling his weight … we're satisfied that he's keeping his mind on the job … yes, in general I should say *Bufo* has more than earned his passage money'.[28]

---

21 Kurth 1942.
22 Pemberton 1934.
23 Pemberton 1932b, p. 138.
24 *The Land*, 30 August 1947.
25 *The Canberra Times*, 29 May 1947.
26 Town Clerk and city manager, Maryborough City Council 1949.
27 Tyler 1976.
28 The Giant Toad 1949.

12 Living with Bufo

Mungomery was deluded.

In the 1950s many towns in rural Queensland were connected to an electricity supply for the first time. In Miriam Vale, a short string of settlement dropped into the bush between the mountains and the sea north of Bundaberg's canefields, there were celebrations when the street lights came. All manner of dignitaries turned up, a brass band marched and billy carts raced along the brightly lit main street. A young barefoot Ian Dahl, his race over, watched as swarms of moths attracted by the lights invaded town, and then cane toads by the thousand attracted by food dropping from the sky.

From day one, toads ignored instructions to eat cane beetles and they foraged instead for ants, caterpillars, slugs and snails. It wasn't their original purpose but it was good marks for the useful toad. And gardeners liked the way they cleaned up garden pests. Gardeners are landscape pests themselves, releasing exotic invaders like *Lantana* and prickly-pear, destroying indigenous colonising plants like native grasses and nitrogen-fixing creepers, and removing caterpillars, slugs and snails that graze their precious trophies. In Hawai'i, toads were appreciated for cleaning up garden and household pests and, just two years after toads were released in Queensland, BSES received 'favourable reports of the toads' work in reducing insect pests in vegetable gardens and household pests such as cockroaches'.[29]

There was no bigger garden in Brisbane than the Botanic Gardens, and the Curator, Ernest Bick,[30] requested some toads, but BSES could spare none from the cane growers' quota. Soon enough they would be hopping all through the gardens uninvited. However, another gardener, Mrs White, got a consignments of toads for her vegetable garden and thought they were 'marvellous'. She was the wife of the lighthouse keeper on Percy Island, offshore from Gladstone in central Queensland. Seven years later she praised the 'simply wonderful [toads]... now, not one [slug or snail] is to be found'[31] but 'vile crows have started to destroy toads by turning them on their backs and eating no other part but the stomach'.[32] Crows are smart.

---

29   *Sydney Morning Herald*. 22 September 1937, p. 15.
30   Pathologist BSES 1935.
31   White 1940.
32   White 1947.

Gardeners in Byron Bay, in northeast New South Wales, introduced cane toads there in around 1965. Toads were introduced to control garden pests while the main invasion front was still heading south from Queensland.[33] The alternate urban myth is that Byron Bay's drug-using community liked and licked toads or smoked dried toad to get a sub-lethal hallucinogenic dose of bufotenine, and that is why toads were taken there.

Byron Bay lies at 28° south, a slightly lower latitude than Bermuda at 32° north; the highest latitude at which toads had previously been liberated. In Bermuda, it was also gardeners. Nathaniel Vesey, a retired Bermudan sea captain, liberated 'about two dozen' of the 'Guianan toads' there in 1875 to eat garden pests.[34]

Byron Bay is not far south of the border with Queensland. Like many borders, people attribute great significance to it, but it was invisible to giant American toads. They crossed it in 1978. In New South Wales, just south of the border, the Condong Mill had been milling sugar on the banks of the Tweed River for almost 100 seasons. In these cane fields, in 1891, visiting American entomologist Albert Koebele recommended importing toads to control beetles in sugar cane.[35] Toads arrived there 87 years later and soon joined up with their cousins a little further south in Byron Bay.

Today, nobody calls these creatures giant American toads or giant toads; they are just called cane toads. It seems to have begun around 1949 when the Maryborough Town Clerk referred to the giant American toad as the cane toad when he complained to the Minister.[36] He was likely not the first to use the name cane toad, but the name stuck. Use of the names giant American toad, giant toad and Queensland toad died out, the common name for *Bufo marinus* became cane toad, and the name cane toad was adopted around the world.

Some came to love cane toads and let their children play with them. Marie Roth of Gordonvale knew two little girls who

---

33   van Beurden & Grigg 1980, pp. 305–10.
34   Waite 1901.
35   Koebele 1891.
36   Town Clerk and city manager, Maryborough City Council 1949.

12 Living with Bufo

> instead of little dollies had these cane toads, they actually had little dresses made up for them, little skirts and beds and a doll's house. They used to dress the toads up and tuck them into their little beds, they used to carry them about ... like you do little dollies. These girls had names for them, they used to set up little tea parties and they used to get these toads and scratch their tummy and the things would lay back obviously enjoying it and stick their legs up into the air and just fall asleep, and they were the most contented ... and I suppose alive little dollies that any girl could like – but just so ugly![37]

Mark Lewis, Director of the 1988 movie *Cane toads: An Unnatural History*[38] filmed Monica Krauss, a young Queensland girl who had a pet cane toad. She called it Dairy Queen after the brand of ice cream container that she used to carry the toad around. Monica would feed the toad mice, help it dance as she sang songs and tickled its tummy when it was tired. And Dairy Queen became the poster toad for the movie.

There were many intriguing characters in Mark Lewis's movie. Elvie Grieg, resident of Redcliffe, a coastal suburb north of Brisbane,

> started feeding toads Whiskettes [cat food] because they started robbing the cat's dish [in the kitchen] ... so I used to put dishes of Whiskettes out and they didn't come inside, so everybody was happy ... they're just friends ... I suppose because they're around the place, you get used to them ... their singing, their calling, it's not only a pleasant noise, it's a friendly noise and I love it. [If] anyone tried to hurt one of my toads ... there'd be a lot of noise and they'd realise I wasn't a lady.[39]

Cane toads became good business for some. From the 1940s there was a profitable trade in live toads. Physiology departments in universities and high school biology classes found them ideal for learning the skill of dissection. Lie the patient etherised upon on a table, toad on its back, arms and legs pinned to the cork board, run a scalpel down the front of the abdomen splitting the belly, fold back the outer skin and you see a

---

37   Lewis 1988.
38   Lewis 1988.
39   Lewis 1988.

Figure 12.1 Monica Krauss with her pet cane toad Dairy Queen, Queensland, 1984. (Arthur Mostead, AM Photography, Canberra. Used with permission.)

colour coded anatomy lesson – red, blue and yellow organs – no blood, no smell, very clean, lovely dissection items.[40]

And cane toads became the creature of choice for human pregnancy tests. It is a tale of innovation. Urine from pregnant women contains gonadotrophic hormones, chemicals that stimulate reproductive organs. In 1928, two German scientists, Aschheim and Zondek, discovered that after injecting urine from pregnant women into mice, physiological changes could be seen in the reproductive organs of the

---

40   Lewis 1989, p. 63.

mice, once the mice had been killed and dissected. The Aschheim-Zondek test became the standard pregnancy test for many years. The drawback was that mice took three days to respond and, on the fourth day, had to be killed and dissected by expert technicians. Baby rabbits responded quicker than mice. After being injected with urine, baby rabbit reproductive organs responded within 36 to 48 hours but they too had to be killed and dissected in order to see the results. Frogs were even quicker. Within 24 hours of being injected with the urine of a pregnant woman, the female South African clawed frog *Xenopus laevis* would lay several hundred eggs. Then a South American researcher, Galli-Mainini, discovered that if a male toad, *Bufo arenarum*, was injected with urine from a pregnant woman, the gonadotrophic hormone in the woman's urine would stimulate the toad to produce spermatozoa in its urine within two or three hours. These sperm could be seen when placed on a slide under a microscope. It was a quantum leap. The turn-around time for a pregnancy test had been brought down from four days to less than four hours, courtesy of toad biology.

For a maternity hospital, if it had to keep animals at all, it was much easier to keep toads than mice, rats or rabbits. And while mice, rats and rabbits had to be killed and dissected, toads could be used repeatedly. Just collect their urine.[41] For Hans Bettinger in the Pathology Department of the Womens' Hospital in Melbourne, the problem was getting a reliable supply of either the South African frog *Xenopus laevis* or the South American toad *Bufo arenarum*. Then Bettinger found a cheap local source of toads – Queensland cane toads. In 1948, the Physiology Department in the University of Melbourne had already organised a regular supply of Queensland-bred *Bufo marinus* for practical dissection classes. Bettinger simply tapped into this resource and adapted the Galli-Mainini pregnancy test for cane toads.

Pregnancy testing with cane toads takes some skill. First thing in the morning, don your white coat, select a syringe with a long needle and draw up 10 millilitres of test urine – donated by a nominally pregnant woman. Second, select your male toad. It is not easy. The male has no external penis and is not easy to identify. It is usually smaller than the female, unless it is a small female, a little more strongly col-

---

41   Bettinger 1950a.

oured, has stronger forearms and a dark patch on the inner surface of the thumb. Grip the male toad behind its forelimbs and it will croak and display the embrace reflex. On the other hand, it may just look at you curiously. Third, once a male toad is selected, restrain the toad face down in the palm of your hand, locked in place with your thumb and forefinger.

Fourth, and this needs practice, take the syringe with the needle facing away from your body and inject the needle just under the skin of the hips on one side of the toad and run the needle carefully under the skin across the back of the hips. With the tip of the needle under the skin on the far side of the hips, making sure it does not exit toad, inject all the sample urine into toad. Withdraw needle, ensure no urine leaks from toad, place toad in covered box and leave to stand in a dark corner for four hours at room temperature.

After lunch, take a glass microscope slide. Beware that once you lift the lid on the box, toad will hop away and urinate at the same time – it is the alarm response. Those first drops of toad urine are precious because, if the woman is pregnant, they should contain toad sperm in response to the hormone present in the woman's urine. Have the slide ready in one hand, lift the lid of the box and quickly put the slide under the toad to collect urine. Place a cover slip over the sample, place the slide under the microscope and examine for toad sperm.

Presence of toad sperm usually, but not always, meant happy baby news for the woman who donated the urine sample – job done.

In 1950 it was a simple and, for the times, rapid test. Bettinger could obtain cane toads easily and 'at a fraction of the cost of *Xenopus* and [they] need, as long as they can sit on a moist rag, no care whatever'.[42] Bettinger employed them once a week and did not bother to feed them for three or four months. By then they were like junkies, starved, emaciated and scarred with needle tracks so they were incinerated[43] – toads sacrificed on the altar of family planning. Pregnant women were grateful, but ignorant of the sacrifice. Cane toads were used in this way from the 1950s until introduction of immunoassay pregnancy tests in the 1970s. But, until superseded, widespread adoption of the method required a regular supply of toads.

---

42 Bettinger 1950b.
43 Bettinger & O'Loughlin 1950.

Apex Clubs – social service organisations – in north Queensland recognised a good fund raising opportunity and banded together to collect and sell toads for both dissection and pregnancy testing. Between 1962 and 1969 they sold and air-freighted 65,000 live toads around Australia. Proceeds from sales, together with grant monies, funded an elderly citizens' home – a Toad Hall. In 1976, some 100,000 toads a year were being used in Australia for teaching and research purposes, and were even exported to France and New Zealand.[44]

And there were private providers as well. Phil Pickersgill in Brisbane sold them to private schools at 80 cents a toad. It was a good market because, in 1981, 'When you pay $2,000 a year in school fees, you expect your kid to have his own toad'[45] – now it's a smart phone and laptop.

June 1974, Darwin, northern Australia. Toads escaped dissection at the high school and went on the lam. Police, wildlife officers and townsfolk joined the hunt, searched storm water drains, burnt off grass, and radio stations played mating calls over the air to attract females. The Northern Territory Government put a $30 bounty on toads and the Darwin Conservation Society put up another $7.50. But four smart toads made it to freedom and were never seen again.[46]

January 1986, Brisbane, Queensland. Toadbusting became a source of income. Julie Halikas, age 10, and Nicole Suffolk, age 11, became toadbusters. For $4 they would clear a suburban garden of toads three times in a week, and return the following week for quality control. The film *Ghostbusters* was a box office hit two years before and lent its name to getting rid of unwanted visitors – paranormal or otherwise; 'there's something slimy in your neighbourhood, who are you going to call – Toad Busters'.[47]

What began as a sorry business for cane growers turned into a good business for Apex clubs and budding entrepreneurs. And there were other opportunities in the cane toad businesses – other ways to make cane toads pay their way.

---

44 Tyler 1976.
45 *The Courier Mail*, 12 November 1981.
46 *Readers Digest*, October 1980; *Time*, 5 August 1974.
47 *The Courier Mail*, 18 January 1986.

1979, Canberra. Toad skins and toxins were the 'next big thing' promoted by the Australian Department of Trade and Resources for Japanese and Chinese markets in traditional medicines; bufotenine just had to be milked from the toads.[48] This 'unique opportunity' comes around once a decade or so.

Jim Terpstra, Canberra entrepreneur, proprietor of Terpstra Toad Farms, contracted with an American company to supply 48,000 toad skins a year for leather handbags. And taxidermists make toad products for sale to tourists: cane toad money pouches, key rings, toad legs, cane toad leather handbags and caps.[49] Toad leather is a very modest export earner. And iconic.

Wednesday 29 July 1981, St Paul's Cathedral, London. Prince Charles married Diana Spencer. Their wedding gifts were scrutinised by Rear Admiral Sir Hugh Janion, 'a jovial and rather irreverent naval officer'.[50] He would have been amused by the gift from the Australian Defence Department. It was a book bound in cane toad leather. Fortunately, the tanning process removed the toxins, otherwise the heir to the throne may not have outlived his wife. The toad skin book joined other rare gifts: a silver plated mousetrap, a tonne of Somerset peat, and two cases of specially blended scotch whisky.[51]

In the 1990s, Townsville taxidermist, John Kreuger, usually paid 50 cents for a good sized toad but once paid $50 for a toad from the Townsville rubbish tip. It was 30 centimetres (a foot) long and weighed two kilograms (four pounds).[52] He stuffed these toads and mounted them on wooden stands in human-like poses: carrying a cricket bat, in the 'missionary position' and, looking every inch a Queenslander, a pot-bellied toad holding a can of beer.

Kevin Ladinsky in Proserpine, north Queensland, went further. He stuffed and mounted toads in realistic dioramas: a four man toad tag-team in a wrestling ring, toads in an Australian Rules football game, convict toads wearing leg irons, and a car smash with toad crash victims, toad ambulance officers and toad police. It was hours of labour for

---

48 Anon. 1979.
49 Marino Leather Exports (n.d.).
50 *The Independent.* 23 August 1994.
51 *The New York Times,* 27 July 1981.
52 *Sydney Morning Herald.* 19 April 1996.

works of art. And Kev 'started outing them in [the] Proserpine Show then and ... won first prize and second prize and got best exhibits and all that for them'.[53] That was the start of Kev's Travelling Toad Show. But Queenslanders were sick of cane toads and few came to see them.

Soon, many more Australians were sick of cane toads but had to learn to live with them, learn to make the most of cane toads: popping, dressing, playing with, feeding, dissecting, injecting, collecting, skinning, stuffing, mounting and selling them.

Although the seething mass of toads at the invasion front was shocking – locust swarms were a fair comparison – gardeners were grateful for the pest cleaning service. But there was a dark side. Native Australian fauna, unprepared for the toxic amphibian, took the brunt of the invasion, suffered the spreading toxic toad slick that could eventually cover two million hectares of the Australian continent.[54] Australia's native fauna became frontline troops in the unfolding cane toad wars.

---

53   Ladinsky 2010.
54   Phillips et al. 2003.

13
Cane toad wars

Ever since the invasion began in 1935, Australia's fauna has been at war with cane toads. Toads invaded with primitive defences, replaced frontline troops rapidly, tolerated harsh environments, and were very effective invaders. For prey, like insects, toads are simply an army on the march, taking all in its path. But for predators, toads are like landmines, squatting, apparently innocent, can be lifted gently without risk, but fatal if mouthed. And a female toad, like a cluster bomb, creates thousands of toxic offspring that hop away into the landscape and lie primed, ready for native fauna unused to toxic toads.

Toads are unsophisticated ordnance.

Some toad invasions were unsuccessful. Toads released in Louisiana died out and toads introduced to Florida spread slowly and had little impact – their potency muted because local predatory birds, snakes and fish evolved together with other poisonous anurans, knew how to eat them and avoid toxic glands, or were immune to the poisons.[1]

Not so in Australia. Australia's isolation over the last 120 million years made it an ark of post-Gondwanan fauna. Some Australian frogs were poisonous to eat but toads were absent. When toads arrived, their toxins were novel. Toxins in Australian frogs are derived from pep-

---

1   Lever 2001, p. 60.

tides whereas toxins in cane toads are derived from steroids.[2] Many iconic Australian fauna such as kangaroo, wallaby and koala eat only vegetation and were not at risk. Many Australian birds had genetic interchange with Asian birds that evolved with toads,[3] or learned to selectively eat non-toxic parts of toads and so avoided the risk. But smaller carnivorous marsupials, cat-sized quoll, mouse-sized dunnart, and carnivorous reptiles, snakes, goannas and freshwater crocodiles were at risk. They were most vulnerable to toads invading their habitats.

In 1935 most thought it preposterous that cane toads, amphibians dependent on water, could invade much of Australia's harsh outback country, but they did. They fanned out from invasion cells planted by BSES in cane farms along Australia's east coast.

Unlike mankind's wars, no one has kept a count of the fallen in the cane toad wars. From the perspective of an ant or a native bee, a dunnart or a quoll, a snake or a goanna, the war has cost innumerable lives. Although the death of ants may not seem tragic, the loss of populations of insects at the toads' invasion front threatens the food chain of the ecosystem the insects support. The loss of cohorts of insects presages the loss of more visible fauna and flora from the landscape.

Others, like snakes, died from direct ingestion of toad poison. In Gordonvale, as the evening sun silhouetted the line of the tableland, a red-bellied black snake, *Pseudechis porphyriacus*, made its way along the edge of the cane field looking for rats and mice. It was almost two metres long, sleek and silent, glossy black with its lower flanks crimson red. A black thread of tongue flickered the air ahead of a narrow, neat head. But instead of a rat, the snake came face to face with a large and unmoving toad, an easy prey. The snake watched for movement, a clue to which way the prey would bolt, but the toad propped motionless on its front legs, blinking. The snake reared and in one fluid sinuous movement struck the toad. The snake's muscular length worked like a whip, the momentum driving its small open jaws over the head and shoulders of the toad as its fangs injected a fatal dose of venom. Within seconds, the toad was dead. With the toad in its grasp, the snake paused, twisted, coiled, rolled onto its back showing its cream belly scales, coiled and uncoiled again, slowly this time, then lay still. Dead. Snake and toad

---

2   Shine 2010.
3   Beckman & Shine 2009.

locked together in a fatal embrace. It was the strangest thing the cane farmer had seen when he found them next morning. A poisonous red-bellied black snake killed by a toad.

Leaving cane farms, toads marched westwards up the Great Dividing Range – a smaller challenge than the name suggests. They hopped, they shuffled, they ran on all fours. By using country roads they avoided the moist closed forests of the scarp and soon reached the dairies, tobacco fields and mango farms on the rich volcanic soils of the Atherton Tableland. Toads had been given a head start. In October 1936, Mungomery gave an advance party of toads to Mr Atherton, a farmer on the Tableland.[4] From the Tableland, it was a downhill run westwards onto vast coastal blacksoil plains of savannah woodlands and seasonal wetlands of the Gulf of Carpentaria.

Max Burns, a tank-sinker, dug dams, called water tanks, in the blacksoil plains of the Gulf to water sheep and cattle.[5] He brought the first diesel-powered bulldozers to Julia Creek in western Queensland in 1947. Before then, tank sinking had been a slow job, powered by horse teams. Burns changed the landscape, sinking big tanks for graziers. Many tanks held water from bores sunk to raise deep artesian water and running all year round to water cattle, allowing them to graze the arid landscape of spinifex and sparse acacia woodland. In this tough environment that few believed toads could conquer, water tanks became oases in the blacksoil plains – toad havens, refuges where they could wait-out dry seasons. Where they could crawl into deep moist cracks in the black cracking clays and wait for rain, staging posts for an invasion force and a breeding ground for reinforcements.

In January 1953, water and electricity reached the settlement of Julia Creek in Western Queensland. Rain broke the drought on hot, blacksoil plains 600 km (375 miles) west of the cane fields. Drenching rains fed the river systems that drained north into the Gulf of Carpentaria. Shirley Eckford and her husband Jim raised cattle on Isobel Downs, just east of Julia Creek. They had their own electricity generator, shedding pools of soft pulsating light around the homestead at night. It was a magnet for cane toads; they 'arrived like troops, an invading army'.[6] When Shirley's domestic helper first saw one it was so

---

4 Bell 1936.
5 Burns 2010.

big she thought it was a bantam hen. Toads re-hydrated and bred in the bore drains. They bred on the home dam that served as a swimming pool, calling out all night and driving the family to distraction. And all the stately goannas around the homestead died when they encountered toads.

Graziers soon became concerned about toads polluting the tanks and bore drains, and being responsible for deaths of cattle – unexplained stock losses. The United Graziers Association enlisted the help of the Zoology Department of the University of Queensland to survey their members. From 1,500 circulars sent out, they received 500 replies of which 123 reported cane toads in their districts. Of these, only nine graziers reported pollution of water supplies by dead toads and only three believed that cane toads had caused stock losses. Instead, graziers reported deaths of working dogs from eating toads, deaths of poultry after their water was polluted by toads, and deaths of native animals after coming in contact with toads – snakes, goannas, fish, eels and birds including water fowl and kookaburra.[7]

The catastrophe for native fauna was unfolding in slow motion.

When both rain and electricity reached the settlement of Julia Creek, insects swarmed at night in yellow haloes under streetlamps, frantic, making the most of the good times. All this food was heaven for the native northern quoll *Dasyurus hallucatus*. These little grey-brown white-spotted marsupial cat-like animals hunt aggressively at night and catch anything edible on the move. A quoll, nose twitching at the end of a long snout, big ears erect, large eyes alert for movement, sensed a large prey, squatting, waiting with others near the street lamp, catching bugs. It was a toad, new to the quoll's territory. The quoll darted, attacked the toad head on, bit deep and fiercely, crushing the toad's skull, ejecting sticky milky poison into its own mouth. In less than a minute the quoll's mouth was foaming, it thrashed the ground, eyes rolled back, tail twisted into knots, tiny body wracked with spasms. It died. A mother, milk glands full for her six offspring. They, blind in her pouch, in turn died slowly from dehydration.

Julia Creek is home to the Julia Creek dunnart *Sminthopsis douglasi*, a small mouse-like marsupial in the same family as quolls

---

6 Eckford 2012.
7 Lee & Straughan 1961.

– dasyurids. Like quolls, dunnarts are nocturnal hunters but small enough for a cane toad to eat – bite-sized marsupials. When cane toads found dunnarts' nests, they camped outside and waited for fast food. Dunnarts, like quolls, attacked small toads and died, but they were quick learners; they learned to avoid toads if they survived the first encounter.[8]

The ground-nesting rainbow bee-eater *Merops ornatus* suffered. They provide spectacular colour in a dusty brown country – brilliantly coloured, green body with golden crown, black mask around the eyes, bright blue cheeks, black and gold throat, bright blue rump and long black tail feathers. Flocks of them feed on insects on the wing and they nest in small burrows in stream banks. Cane toads found them, ate their eggs, ate their chicks, made homes in the burrows and enjoyed their nests.[9]

In 1969, CSIRO began introducing dung beetles to bury cattle dung and thereby remove breeding sites for flies that were pest of cattle. But in 1972 Queensland cattlemen noticed that cane toad were eating precious dung beetles, and the Muttaburra branch of the Australian Country Party demanded of the Queensland Government that something be done about the cane toad menace.[10] BSES Director, Norman King, advised the Director General of the Department of Primary Industries that 'it would be appropriate to point out that the toads were here for 35 years before the dung beetle was introduced'.[11] CSIRO joined in – the Chief of CSIRO Division of Entomology, Doug Waterhouse, alerted Norman King to the possibility that *Bufo marinus* was an obstacle to the success of the dung beetle program and asked what the current view was of the significance of the toad to the sugar industry. Tellingly, Norman King replied, 'It is the view of our entomologists that the cane toad is of no significant value in controlling insects injurious to cane.'[12] But he offered no respite for dung beetles – the newcomers on the block. Much later, research confirmed that cane toads likely reduced the effectiveness of the dung beetle program.[13]

---

8   Shine (n.d.).
9   Boland 2004.
10 Secretary Australian Country Party, Queensland 1972.
11 Director BSES 1972.
12 Chief of the Division of Entomology, CSIRO 1972.

In April 1975, cane toads arrived at Boodjamulla, a sacred place, Rainbow Serpent country of the Waanyi people, 300 kilometres west of Mt Isa. Toads walked in and helped themselves. The little rain that falls on sandstone hills and limestone plateaux drains into Lawn Creek catchment, filling the deep gorge of Boodjamulla National Park.[14] Waanyi believe if you pollute the water or take it for granted the Rainbow Serpent will leave and take all the water away. Quolls, freshwater crocodiles, goannas and snakes died resisting invaders. Boodjamulla holds the Riversleigh fossil beds of extinct turtles, crocodiles, pythons, birds, thylacines, giant wombats and kangaroos dating back 25 million years. These fossilised creatures are now extinct or evolved into today's relatives, but the cane toad hopping around these fossil beds is little changed from how it was back in South America: survivor, invader, irresistible coloniser.

1975, 40 years after cane toads were released in Australia, Mike Archer, mammal specialist from the Queensland Museum, felt the impact of cane toads when his pet quoll died in 'tetanic contractions' after catching a toad. He had tried to save his pet, and 'in 20 minutes that evening I got to three vets in ... Brisbane. Two of the three didn't even know toads were toxic'.[15] Archer's public announcement, that a cane toad was the cause of the death of his pet, stirred old prejudices. William McDougall, a former BSES Assistant Entomologist in the late 1930s and, later, Chief Entomologist of Queensland, confronted Archer in the Queensland Museum, 'jabbed a big angry finger into my chest and growled "I don't know what killed your Quoll, but it wasn't a cane toad"!'[16] When early dog deaths were reported following the release of cane toads, BSES proclaimed the toad's innocence, but in January 1938 Bill Kerr had asked William McDougall to incorporate toad toxicology studies into his program on rats.[17] McDougall told Archer that he had extracted poison from the toads' paratoid glands, dried it, rehydrated it and injected it into the circulatory system of rats where it had no effect. But this is not how the poison works: cane toads have no fangs with

---

13 Gonzalez-Bernal et al. 2013.
14 Queensland Government (n.d.).
15 Archer 2013.
16 Archer 2013.
17 McDougall 1938.

which to inject poison – the poison is passively ingested into the predator through the mouth and digestive tract.

Poor science perpetuated the toad myth. It is true, but hard to believe that, 40 years after cane toads were released in Queensland, the Chief Entomologist of Queensland and many veterinarians still did not understand the mechanism by which would-be predators of cane toads ingested poison and died.

Mike Archer's loss set him on revenge. One night, while out spotting wildlife, he came across a toad, the creature that had killed his pet. He swung his geology hammer and hit the toad on the head, killing it instantly. But he felt excruciating pain. His hammer splattered poison from the toad's parotoid glands across his face and it was six hours before the pain went away and he was able to use the eye again.[18] And there were other troubling incidents: 'It was quite common to have kookaburras [birds] … brought in [to the Queensland Museum] with a cane toad half down it's throat … half way through the process of swallowing this cane toad it falls out of the tree stone dead … just like it had swallowed a cyanide pill.'[19] Archer collaborated with snake specialist Jeanette Covacevich to survey the impact of cane toads on native fauna.[20] They interviewed veterinarians and other scientists and sent thousands of circulars out to schools and National Parks offices in the parts of Queensland occupied by toads. Results were clear. Toads were fatal for more than 11 native animals, from goannas to snakes, birds and quolls, and even a Tasmanian devil, *Sarcophilus harrisii*, captive in Queensland, that died after mouthing a cane toad. It was a sobering assessment and, 40 years after cane toads were introduced, the first review of their impact on native fauna to be published in a scientific journal.

In December 1979, cane toads invaded the vast flat savannah country of the Gulf of Carpentaria, northern Australia – a landscape of contradictions, both giving and unforgiving. Toads crossed the Bynoe, Flinders and Leichhardt rivers,[21] once believed impassable for toads. They advanced in the wettest months from December to March, when rains greened the landscape, their advance slowed by man-high grasses.

---

18   Lewis 1988.
19   Lewis 1988.
20   Covacevich & Archer 1975.
21   *The Courier Mail*, 14 May 1979.

And toads drowned when waters flooded the gently sloping black soil plains. When floods receded, many more toads died from desiccation during eight months of the dry season; died on outcrops of bare broken rock, on skeletal soils supporting vegetation inured to tough times. Resinous grasses miserly with their proteins, and stocky, leathery leaved eucalypt and acacia trees survived by dint of a few deep searching roots. Toads died in suffocating pools of fine silt – bulldust; other toads survived 'the dry' by camping at waterholes and stock tanks, hiding in deep fissures in black, cracking clay soils, in dead tree trunks or moist clefts in rocky outcrops. When rains came once more, fecund females spawning 30,000 eggs at a time bred the storm troopers of a reinvigorated invasion force.

In April 1982, cane toads invaded Australia's Northern Territory. There was no resistance. They invaded westwards at 27 km/yr (17 miles/yr).[22] At the invasion front there was one toad for every 5 square metres (4 square yards). Bill Freeland, wildlife ecologist with the Conservation Commission of the Northern Territory, warned that toads would eat native fauna, poison native animals that tried to eat toads, compete with them for food, shelter and resting places, and compete with native amphibians for breeding habitats. They were an economic and social liability.[23]

In 1988, cane toads reached the small township of Borroloola, 1,600 km (1,000 miles) west of the cane fields, completing an epic march to the western end of the Gulf. Brendan Wilson ran the pub at Borroloola. He was used to unruly characters. Every morning there were around 30 uninvited guests in the swimming pool. Cane toads couldn't get out once they were in. Every morning he fished them out and bashed each one to death against the fence. But more toads were always there next morning.[24]

Graeme Dingwall ran fishing tours at Borroloola. Cane toads did not seem to bother barramundi fish or saltwater crocodiles in the creeks and estuaries around Borroloola. But one of his inquisitive young dogs mouthed a cane toad, went very quiet, hung its head and began to vomit. Dingwall, sitting under a mango tree drinking beer,

---

22 Freeland 1986.
23 Freeland 1985.
24 Begg et al. 2002, p. 8.

worried about the pup, then opened another beer. Dog sick, owner depressed – so he opened another beer or two. Late at night the pup got up gingerly, wagged its tail and tottered off to its bed; its owner opened another beer to celebrate the miraculous recovery, and another beer for good measure. In the morning the pup, bright eyed, mouth open, tongue hanging out, was ready to play, but now the owner was as sick as a dog.[25]

If you lived in town, toads were an inconvenience, but if you were an Indigenous Australian who lived in the bush and collected wild food, toads were a disaster. At Robinson River Aboriginal community, there was no more sugarbag honey because toads had eaten native bees. There were no more delicious fat goannas to roast on the fire because goannas mouthed toads and died – the same for snakes and lizards.[26] If bush tucker goes, people go too.

In 1990, 55 years after Sir David Rivett of CSIR supported release of toads, Hugh Tyndale Biscoe of its successor CSIRO remarked, 'It is astonishing that so little is known about the biology of the [cane] toad.'[27] The remark was timed to coincide with the Australian Government's grant of $1.25 million to CSIRO to research controls for cane toads using a virus or a bacterium. Viruses were in CSIRO's pedigree, founded on their partial success in 1950 with myxomatosis, a virus that almost eliminated Australia's feral rabbit populations. CSIRO's carefully controlled release of the virus got out of hand when farmers simply collected infected sick rabbits and released them on their own properties with no controls. Large numbers of feral rabbits died very quickly but, largely because of the loss of control over the spread of the disease, Australia's rabbit population gained immunity and bounced back.[28] In the case of anurans, viruses are an exceptionally difficult and risky form of biological control because of the dangers of a toad virus spreading to other amphibians. CSIRO's toad virus program and later research into biological control of cane toads using molecular biology drew a blank.[29] Thankfully, Australia was spared another disaster in biological control.

---

25   Legislative Assembly of the Northern Territory 2003, p. 6.
26   Legislative Assembly of the Northern Territory 2003, p. 6.
27   *New Scientist*, 24 March 1990, p. 5.
28   Rolls 1969.
29   Shanmuganathan et al. 2010.

In 1995, 60 years after the cane toad invasion began, CSIRO[30] began a detailed survey of the impact of cane toads. Over two years, the team measured the impact on wildlife as the invasion front advanced through the Roper River region at the base of the Top End of the Northern Territory. They chose three habitats: the toad invasion front, the east where toads had been present for some time, and the west where toads had not reached. There, they surveyed an abundance of creatures: 25 invertebrates, including spiders, ants and beetles; 14 amphibians (mainly frogs); 47 reptiles including crocodiles, snakes, lizards and goannas; 171 bird species; and 17 mammals including dunnart, wallaby, buffalo and dingo. They surveyed the sites both before and after the end of the wet season (October and May) in each of two years. It was an extraordinary effort.

All three sites were next to billabongs, deep pools in rivers in the eucalypt savannah woodland of the western Gulf country. On the banks, solitary thick trunks of ancient river red gums support spreading crowns, and groups of pale paperbarks grow close to water. Away from water, sparse and wiry grassland is dotted with pale trunks of ghost gums and brown fibrous trunks of coolibah and bloodwoods. It was a hot and dry landscape between billabongs – tough going for amphibians.

Once the data were in, CSIRO announced to astonished readers that there was 'little evidence that cane toads have a significant adverse effect on the diversity and abundance of many of the native fauna examined'.[31] According to them, cane toads threatened only three native animals: the dingo *Canis lupus dingo*, Roth's tree frog *Litoria rothii*, and Gilbert's dragon *Lophognathus gilberti*. But all three animals had also been found happily sharing habitats with cane toads in other areas.

Australia's peak science body declared that cane toads were not a problem for native wildlife. The evidence was circumstantial; cane toads were found not guilty, and freed without charge.

CSIRO's conclusion was curious, counter intuitive, and contradicted local knowledge. It was science, but did not reflect the experience of people of the Gulf country. Locals could have told them differently – no sugarbag honey, no lizards, no goannas, no quolls, rainbow bee-

---

30   Catling et al. 1999.
31   Catling et al. 1999.

eaters gone from some nests, dead freshwater crocodiles and some sick fish.

No adverse effects of cane toads – statistically proven by scientists! It is all in the experimental design. Firstly, in the Roper River region where the study was done, there were low numbers of several goanna species, and it was outside the range of quolls, so effects of cane toads on at least these two vulnerable animals were impossible to see. Secondly, the study period was short, but natural variations in populations of sparsely spread creatures was very high. This makes differences in populations of native fauna due to cane toads very difficult to confirm because real differences in populations could not be separated from differences that may have happened by chance – the benchmark of statistical tests. To quantify the impact of cane toads on native animals, the study should have been done in areas where animals were plentiful and the study should have run for many more seasons – strategies that would increase sample size and reduce the impact of variation in population sizes with seasons. Better located sites would have become long-term monitoring stations. But it would have cost a great deal more money and would need to have been run by a stable group of researchers with long employment contracts – an organisation just like CSIRO perhaps.

But the research effort was about to get a boost. Small research teams working in the Top End of the Northern Territory of Australia had collected a wealth of background information on key populations of native fauna. By 2005, when cane toads arrived there, scientists were ready. They measured the true impact of the cane toad invasion front.

14
Taking the Top End

The Top End of the Northern Territory of Australia lies between 12° and 15° south of the equator. It is a comfortable climate for amphibians: warm temperatures all year round and plentiful water for breeding in summer.

Sedimentary rocks more than 1,700 million years old form the Arnhem Land plateau of the Top End. These rocks were formed when life was still confined to the sea and the atmosphere was toxic to oxygen breathers, before tetrapods moved onto land, before the advent of amphibians. At its western end, the bold scarp of the plateau keeps watch over coastal Wetlands of International Importance[1] and World Heritage Kakadu National Park. Rock shelters in Kakadu scarp contain some of the oldest evidence of human habitation in Australia – at least 60,000 years old – from when humans migrated there from the north across shallow seas. The first cane toads arrived in Kakadu National Park from the east in January 2001,[2] 66 years after their release in Queensland. It was a journey of around 1,800 km (1,120 miles) from the cane fields.

The Kakadu wetlands are home to magpie geese, duck, jabiru, lotus birds, cormorant, darter, egret, ibis, heron, sea eagle and other raptors. And in the wet season they are home to around 30 species of birds that migrate there. Until recently, Kakadu's wetlands were also home to

---

1   Parks Australia: Kakadu National Park (n.d.).
2   van Dam et al. 2002.

herds of feral water buffalo *Bubalis bubalis*.[3] Buffalo were imported into the Northern Territory from Indonesia in the 19th century to provide working animals and meat for remote northern settlements. They soon went feral and disrupted and polluted much of the wetland ecosystem, and became dangerous pests. In 1978, control of feral buffalo began in earnest for economic reasons – protection of domestic stock from disease. The Northern Territory Government's program to wipe out brucellosis and protect the domestic cattle industry resulted in thousands of buffalo being shot and eradicated from Kakadu National Park. Just as the wetlands began to return to their natural state, cane toads arrived, and took over buffalo wallows, crocodile nesting grounds, bird habitats and the World Heritage Wetland.

There is no eradication program for cane toads.

In April 2001, Beryl Smith, Indigenous Jawoyn person, Kakadu Park Ranger, was getting ready for a night-time fauna survey with fellow Ranger Kathy Wilson, dropped by helicopter onto the sandy banks of Gimbat Creek. It flows into the Katherine River and the gorge of Nitmiluk National Park – cicada dreaming country of the Jawoyn. At dusk, Beryl and Kathy set off with spotlights to count reflections from animals' eyes along the creek. They were in for a shock. 'We walked about 500 metres and found around thirty toads in about half an hour – by 8 o'clock dozens of them had started calling … it gave me a feeling of dread.'[4] Cane toads were right through the creek system, males calling to females, toxic toads breeding in sacred country, in World Heritage ecosystems. Beryl was sad: 'Jawoyn country is a beautiful place but I don't think it will ever be the same again.'[5]

Cane toads spread through Nitmiluk Gorge[6] system, right through the base of the Top End. It was catastrophic for native fauna unused to the invaders – fatal for bee, snake, goanna, freshwater crocodile and quoll. And repercussions would flow, predator then prey, removing pieces of the food chain, a catastrophe in slow motion. Those who saw the links could sense the repercussions, could see gaps in the future, were sad, dismayed, distressed at what was to come.

---

3   Northern Territory Department of Land Resource Management (n.d.).
4   *Sydney Morning Herald*, 7 April 2001.
5   *Sydney Morning Herald*, 7 April 2001.
6   Legislative Assembly of the Northern Territory 2003, p. 50.

14 Taking the Top End

In October 2001, cane toads invaded the town of Katherine at the base of the Top End.[7] They found new homes around houses, dogs caught toads and died, and parents worried about their children catching and playing with toads. It was obvious that cane toads would soon reach the Daly River, west of Katherine.

The impact of cane toads on the Daly River would be unequivocal because a team of researchers had been studying the ecosystem. Leading the team was Sean Doody who started on the project as a doctoral student with Professor Georges at the University of Canberra. Sean was raised in Baton Rouge, Louisiana, on the banks of the Mississippi where, as a child, he collected frogs, lizards and anything else he could catch. That led to study, and his master's research on turtles led to the invitation from Professor Georges to work in the Northern Territory. In 1996, Sean joined a three-year program monitoring turtle, goanna, crocodile and wallaby on the Daly River. He spent six months of each year camping with the survey team through the dry season in the remote, vast, Top End savannah. It was paradise for a wildlife ecologist plucked from the banks of the Mississippi. His task was to document the impact of goanna on turtle eggs and turtle populations. The yellow-spotted goanna *Varanus panoptes*, favourite 'bush tucker' food for Indigenous people, lives along the Daly and other rivers in the Top End. It is a large monitor lizard about one and a half metres long, dark brown with bands of large black spots and smaller pale yellow spots. It sometimes stands erect on its back legs, supported by its tail, and monitors its surrounds – hence monitor lizard, commonly called goanna – a corruption of iguana. Goannas forage only during the day and need warmth from the sun to drive their metabolism. Goannas have a remarkable sense of smell, use forked tongues to sense buried food, and a favourite is buried eggs of the pig-nosed turtle, *Carattochelys insculpta*. This turtle lives in fresh water habitats in northern Australia and in southern rivers of New Guinea. It is the only living member of its ancient Gondwanan family, Carettochelyidae; its forebears are preserved as fossils. It is about the size of a large dinner plate, has large eyes and a long nose, flattened at the end, like a pig's nose. Survey completed, the evidence

---

7 Cane toad sightings in Frogwatch (n.d.).

was clear. Goannas dug up around 20% of turtle nests to eat eggs and were the dominant predator of turtles.

The Daly River was a living laboratory. The team could measure the impact on wildlife as cane toads arrived. It was too good an opportunity to pass up. The team continued monitoring, using their own resources for two more years until Australian Government funders woke up. Monitoring was expanded to include other animals at risk: Mitchell's monitor, *Varanus mitchellii*, that forages in pandanus trees, Merten's water monitor, *Varanus mertensi*, that forages around river banks, and Gilbert's dragon, *Amphibolurus gilberti*, a small lizard preyed on by goannas. And the team continued to monitor freshwater crocodiles, *Crocodylus johnstoni*, and pig-nosed turtles. The team's continuous seven-year ecological record of the Daly River ecosystem provided the foundation of strong science, underpinning well-documented conclusions about the impact of invading cane toads.

In the tropical wet season of late 2003, cane toads reached the Daly riverbank. The impact on the river system was profound. Large yellow-spotted monitor lizards that roam the landscape were first to find toads, and first to die. They just had to mouth a toad and they were dead. Populations of goanna along the river bank crashed. Up to 97% of goannas – almost all of them – died.[8] Freshwater crocodiles died as well, took easy meals, snacked on cane toads, and died with toxic toads in their stomachs. They filled with gasses of decay, rolled onto their backs and floated downstream, white bellies reflecting the hot sun. And tiny quolls wiped themselves out, vanished from the nocturnal landscape. Cane toads were fatal for quolls. Before the invasion, quolls were regular evening visitors to the river bank. But after cane toads arrived, the only rustling in the evening was toads looking for food.

It was not all bad news. Nature adjusted, species responded to dead predators, to liberation. Gilbert's dragon was one. Its main predator, the yellow-spotted monitor, had been removed from the food chain, so the population of water dragons increased. And pig-nosed turtles had a reprieve: because the yellow-spotted monitor had been removed from the food chain, turtle eggs were left alone in their buried nests.[9] And Sean Doody believes populations of finches and wrens that nest in pandanus

---

8   Doody et al. 2009.
9   Doody et al. 2006.

trees also got a reprieve because of the sudden deaths of Mitchell's monitors that ate cane toads as a first course before their favourite birds' eggs.

A similar repository of pre-invasion ecology had been created near the north coast of the Top End at Fogg Dam, a shallow lagoon – relic of a failed rice project – on the flood plain of the Adelaide River, east of Darwin. Professor Rick Shine, leading a team of zoologists and ecologists from the University of Sydney, had been studying the snakes of the flood plain since 1985. He had gathered 20 years of information on the ecology of the flood plain before cane toads arrived there.

In 2005, cane toads hopping west from Kakadu crossed the Adelaide River and reached the University of Sydney researchers – renamed Team Bufo – at Fogg Dam. A wealth of new information on the impact of cane toad invasions on native fauna and on the active and rapid evolution of toad populations would come out of this concentration of scientific brains. Because of Shine's predilection for snakes, one of the first studies was on the impact of cane toads on snakes. Ben Phillips was the research student. Ben grew up chasing snakes in northeast New South Wales and saw, first hand, the effects of cane toads on red-bellied black snakes. Shine had been conducting research in central Australia and found Ben searching a rubbish dump near Alice Springs for Bynoe's gecko *Heteronotia binoei*, small lizards that live in all-female populations and don't need males to breed. This chance meeting led to a scholarship for Ben to study cane toads with Shine and an opportunity to work with snakes and cane toads. Back in 1970, in the name of science, Richard Wassersug had fed cane toad tadpoles to human colleagues.[10] Some thirty years later, Ben fed cane toad poisons to Australian snakes – with clearance from the University of Sydney's ethics committee. Inspired research is a constant, but ethics committees now scrutinise research proposals. A proposal to feed tadpoles to humans would not even reach the committee now – a reflection on how the practice of science changes over time.

Ben required a kitchen blender, a paddling pool, 78 cane toads and 135 snakes.[11] Cane toads were skinned and skins went into the blender to make essence of toad. Snakes went into the paddling pool,

---

10 Wassersug 1971; Wassersug 2008.
11 Phillips et al. 2003.

Figure 14.1 Ben Phillips. (Ben Phillips. Used with permission.)

first for a dip to see how fast they could swim, and again after being force-fed essence of toad through a syringe. Unexpectedly, some snakes died immediately. Most could tolerate only a small amount of blended toad skins. It was a shock. Even small cane toads contain considerable amounts of toxin and most snakes, despite their slender shape, can swallow a surprisingly large prey for dinner. Snakes were very susceptible to toad toxins. But two species of snake could take a shot of toad and survive: the keelback snake *Tropidonophis mairii* and the slatey-grey snake *Stegonotus cucullatus*. The research team's message was grim for others: 'The invasion of cane toads is likely to have caused, and will continue to cause, massive mortality among snakes in Australia.'[12] It was a very different conclusion from CSIRO's, published just four years earlier. But even this simple conclusion from a controlled experiment would prove presumptuous. Its premise about close interactions

---

12 Phillips et al. 2003.

between populations of snakes and toads was not generally applicable. Deaths of goannas from eating cane toads benefits snake populations more than the deaths of snakes from eating toads reduces snake populations[13] – for snakes, 'the enemy (cane toad) of my enemy (goanna) is my friend'.

Ben and his colleagues also fed cane toad essence to other high level Australian predators:[14] 11 species of reptiles including a python, freshwater turtle, two species of crocodile, two species of dragon and five species of goanna. Again, the speed of each animal was clocked before and after being fed with essence of toad. The results were predictable. All species except crocodiles were very sensitive to toad toxins and would die if they ate toads. Freshwater crocodiles (freshies) died and only saltwater crocodiles (salties) could tolerate doses of toad toxins. It confirmed what locals knew. Predators that mouthed a toad would die, except for a couple of snakes, and salties would leave toads alone on the river bank because toads made them sick.

Later, Team Bufo discovered that although many vulnerable species of snakes will die from attacking toads, in some species some individuals will not eat toads and will survive. These individuals with different hunting habits, or perhaps mouths too small to eat a toad, build new populations of snakes immune to the presence of cane toads. But some snakes will never be affected by cane toads. Keelback snakes, common at Fogg Dam, are of Asian lineage and can eat toads without any problems. And although the olive python *Liasis olivaceus* is very susceptible to cane toad poison, it seems not to recognise cane toads as food.

Northern Australia's iconic fish the barramundi, *Lates calcarifer*, is a popular subject of ancient, faded cave paintings in the Kakadu scarp and is a modern drawcard for fishers to the Top End. It also proved very discriminating – a survivor. After a couple of attempts to eat a cane toad tadpole, spitting out the bitter wriggler, barramundi ignored them.

Team Bufo thought that tiny planigales, a small dasyurid like the quoll, would be as vulnerable as the quolls themselves. But, after being introduced to a cane toad and, provided they did not die in the first

---

13   Shine 2012.
14   Smith & Phillips 2006.

encounter, planigales ignored toads, avoiding them along with other unpalatable prey.

Some frogs are vulnerable to cane toads because they eat other frogs and so attacked cane toads when they appeared at Fogg Dam. When cane toads arrived there, team Bufo found many dead lilypad frogs, *Litoria dahlii*, floating in the water.[15]

The research at Daly River and Fogg Dam on the Adelaide River floodplain revealed both despair and hope: despair that whole local populations of some iconic Top End fauna could be wiped out, but hope that some species would ignore cane toads as food or learn very quickly not to eat them; hope that some species would now become more numerous because their predators had succumbed to toads; and hope that, for some, a subset of the species – different perhaps in size, shape or food preference – would behave differently from the rest, avoid toads, and breed to regenerate a new population modified in situ to its new conditions.

But the one unavoidable conclusion is that cane toads utterly changed this part of the world.

Research conclusions of this quality from the teams at Daly River and Fogg Dam were possible because of long term pre-invasion records collected over years of pedestrian research. In both places the research was serendipitous – at the outset, neither group planned for cane toads to invade their research habitats – and for that we should be thankful.

In January 2005, the height of the Top End wet season, the toad invasion front was not far from the Northern Territory's coastal capital city, Darwin. Humpty Doo is a rural settlement on Darwin's outskirts – a mix of mango orchards, grazing land and residential two-hectare rural blocks where children play, liberated from bitumen and concrete of suburban streets – the archetypal 'Territory lifestyle'. Cane toads hopped in, liked the lifestyle, killed pet dogs, ate bowls of pet food and made locals fret for children's safety. Locals were not happy: toxic toads were everywhere, calling, breeding, on the roads, in the laundry, in the swimming pool where they couldn't get out, where they drowned, where owners believed the water was poisoned. Owners donned rubber gloves, captured toads, put them in double plastic bags, put them in the

---

15   Shine 2007.

freezer, and froze the buggers to death. But there were more toads than freezer space. The invasion front swept through the rural Territory lifestyle and on to the capital, Darwin.

In 2006 cane toads from Darwin's rural area invaded tropical suburbs, finding homes in golf courses, under outdoor barbecue areas, in carports and among potted palms. They liked backyard swimming pools. They liked the lights on the front porch. They took to the urban lifestyle where food was on hand. They were 'the scourge of the Top End ... the frog from hell'.[16]

Darwin-based Frogwatch NT began toadbusting – toadmustering – when cane toads hit town. It was not a cash earner like in Brisbane in the 1980s; this time it was to stop cane toads overwhelming ecosystems, to protect habitats of native species and protect breeding areas for native amphibians and reptiles.[17] Frogwatch also promoted toad traps and toad-proof fences to keep toads out of sensitive areas. It was the first tangible resistance toads confronted on their long march west. Toadmusters were popular, where people got together at night to remove all visible toads from an area, then destroy them. Adults and children were quite happy to catch, bag and kill toads – death by gassing with carbon dioxide, by spraying with disinfectant, by smearing with haemorrhoid cream, by freezing, or bashing to death. If asked to kill feral cats that catch and kill wildlife, asked to gas, freeze or bash feral moggies, the same people would be horrified and children would not be allowed to do it – yet they killed toads with relish. Humans, the most irrational of rational beings, caught and destroyed 58,000 toads in five years[18] – the annual offspring of perhaps one or two female toads – making little impact on toad populations around Darwin but satisfying participants who believed they made a difference by protecting small populations of native fauna.

Toadbusting is akin to picking up cane grubs from cane fields – even collecting two tons of cane grubs made no difference to the pest. It is ineffective but it makes the participants feel good, feel they are doing something, protecting their patch. Predictably, despite all the good in-

---

16   *Sydney Morning Herald*, 17 September 2007, p. 16.
17   Frogwatch (n.d.).
18   Frogwatch (n.d.).

tentions, toads overran Darwin and its suburbs, entrenched themselves in the urban landscape and dug in for the duration.

As the northern offshoot of the invasion hit Darwin at the tip of the Top End, the western front hit the lower reaches of the Victoria River at the western base of the Top End. A stretch of the Victoria River, 100 km (60 miles) upstream from its mouth, was named Long Reach by Lieutenant Stokes of the *Beagle* when in 1839 he 'disturbed the stillness that had reigned there for years' by shooting a crocodile. 'It was not before he had received six [musket] balls in the head, that he consented to be killed.'[19] It is an oasis of sorts in a dry, bleached landscape. When cane toads reached Long Reach (now known as Longreach Lagoon) in the dry season of 2007 there was mass mortality of freshwater crocodiles.

Mike Letnic, wildlife ecologist who monitored crocodiles for Northern Territory Parks and Wildlife, got reports of mass deaths from helicopter pilots who flew over the Victoria River. He did a night-time spotlight survey to check what was happening and saw freshwater crocodiles catching toads on river banks,[20] then found dead freshies with cane toads in their stomachs. The scene was shocking. Letnic counted 13 dead freshies lying belly up in the river. The smell of decay was overwhelming. As the invasion spread further upstream, the scene was just as bad; dead freshies were everywhere. It was acute because when cane toads arrived at Victoria River, having hopped through the surrounding arid landscape, they congregated in large numbers at the river's edge to rehydrate. And freshies tucked-in to handy toad snacks, the most abundant food. Young adult freshies died in greatest numbers. Death took hours, stomachs filled with gas, carcasses rolled onto their backs, drifted downstream, white belly scales reflecting sunlight, easy to spot. Almost 80% of freshies in the upper reaches of the Victoria River died after toads arrived.[21]

From Victoria River it is 160 km (100 miles) to Western Australia's state border. Since 2003, the founders of Kimberley Toadbusters (KTB)[22] had been preparing to stop cane toads entering Western Australia. A small dedicated group of volunteers harassed the vanguard

---

19 Stokes 2004(1846).
20 Letnic & Ward 2005.
21 Letnic et al. 2008.
22 Kimberley Toadbusters (n.d.).

of cane toads as they approached Kununurra. They used traps, toad-musters where they gassed toads to death, and fenced water sources in the dry season to keep toads in the open where they would dehydrate and die. And it worked – for the toadbusters. It felt good to catch and kill toads. In April 2012, KTB claimed their program had nabbed approximately 2.4 million adult toads[23] but toads still marched westwards in unstoppable numbers.

In January 2009, toads marched on, ran across the West Australian border, and into the Kimberley region. Toads were too numerous, the country too vast, people too few.

In March 2010, in Kununurra, Western Australia, cane toads reached the town at the centre of the Ord River irrigation scheme, an engineered and irrigated landscape at odds with its surrounds – as alien a landscape as cane toads are alien fauna. Argyle Dam, completed in 1971, holds back Lake Argyle, covering 100,000 ha (247,000 acres) and some 10 million megalitres of water. It is a man-made wetland of international significance, Ramsar Site 478. The irrigation scheme provides water to 11,700 ha (29,000 acres) of manicured cropland with shallow pools of fresh water changed regularly. It is cane toad heaven. And heaven is set to get bigger. The irrigation scheme is to be enlarged by an additional 44,000 ha (109,000 acres) – a proudly man-made toad-friendly beautiful wetland set in a spectacular but harsh and unforgiving landscape.

By late 2010, cane toads had overrun Kununurra and were headed towards the Western Kimberley Ranges near the westernmost rim of the continent. Its unique arid habitats isolated from the world by harsh surrounds and scarce surface water are home to ancient relict flora and fauna – landscapes worth protecting. But how?

At best, only small ecological islands can be protected from invasion using barriers at ground level that even small toad hatchlings can't penetrate. One such barrier has been built to protect Emma Gorge in the El Questro Wilderness Park by Stop The Toad Foundation.[24] But in a harsh landscape of droughts, fires and floods, such as the western Kimberley region, maintaining such barriers is a big task. Odds are on the toads to make a clean sweep of northern Australia.

---

23 Kimberley Toadbusters 2012.
24 Stop the toad (n.d.).

There is a sequel to this story that Cyril Pemberton of HSPA would not have been at all proud of. In 2008, toads heading westwards from the Top End were not so chipper, lost weight, became lethargic. They were infected with a parasite, the lungworm *Rhabdias pseudosphaerocephala*. The origin of this lungworm is South America,[25] and the likely mode of transport was in the lungs of toads brought from Hawai'i. It is likely to have been present in toads in Hawai'i prior to their importation to Australia and, by inference, must have been carried in Pemberton's toads from Puerto Rico, and in toads all the way back to those that went from Suriname to Martinique in the 1840s.

The incidence of the parasite in toads in Hawai'i is reportedly much lower than in Australia. This could well be because in 1932 and 1933 at the HSPA facility in Waipio, Denison bred toads from eggs collected in the rice fields. The offspring from these eggs were reared in concrete ponds, remote from their parasite-carrying parents. These toads, distributed around the islands, would likely have been free of parasites. But some of the first batch of toads received at HSPA were released in the arboretum behind Manoa, likely complete with a complement of parasites. When, in 1935, Mungomery, Pemberton and Denison were snatching toads for Queensland, they collected the offspring of some of the Manoa toads – probably together with lungworms. At Meringa in north Queensland, toads were bred naturally. In the breeding pen at Meringa, if the parasite was present it would have moved freely from adult to juvenile toads. All toads may have become infected. If this was the case, the parasite would have been carried into the wild in Australia by the original population of toads released by Mungomery.

An alternative explanation of the presence of the lungworm in cane toads in Australia and Hawai'i is that it arrived in both places on another carrier. But reduced genetic diversity among cane toads in Australia compared to South America is mirrored in lower genetic diversity in lungworms between the two places, and this indicates a similar pathway of introduction.

Throughout their careers, Cyril Pemberton and his HSPA forebears went to great pains to ensure that the insects that they sent back to Hawai'i were free of parasites. Prior to shipping, they would rear suc-

---

25 Dubey & Shine 2008.

cessive generations of the insects to make sure that they were not carrying any hidden baggage. Similarly, albeit as an afterthought, in 1932 Pemberton instructed his HSPA colleagues to check the Puerto Rico toad shipments for ticks and to burn the packing materials. But such was the urgency to get the toads to Hawai'i that he checked neither the efficacy of the toads first-hand, nor whether they carried internal parasites. This was an uncharacteristic lapse in Pemberton's scientific protocols.

Cane toads, together with lungworm parasites, are in Australia. The good news is that, until now, South American lungworms have not infected native Australian frogs, but neither have native Australian lungworms infected cane toads.[26]

Cane toads will inevitably make it to the west coast of the Australian continent. More than 2,000 km (1,240 miles) in little more than 80 years across harsh and unforgiving landscapes is a remarkable journey for an amphibian. As they adapt, subtle but significant evolutionary changes are taking place. In little more than three human generations toads have covered new ground at speeds few thought possible. They appeared in Darwin much faster than anyone expected. In the 1930s toads advanced westwards from coastal cane farms at around 10 km/year and, in 1985, Bill Freeland[27] estimated toads would reach Darwin by 2027 – they arrived there 21 years early in 2006. In the 1950s and 1960s they invaded the Gulf country at 27 km/year. But the invasion front advanced on Darwin at 55 km/year.

Is this evolution in action?

Ben Phillips went back along the path of invasion and discovered that a modest 9% increase in leg length and a three-fold increase in stamina led to an almost six-fold increase in the speed of invasion.[28] The compounding effect was remarkable. He called this the 'Olympic Village' effect – all the best physical specimens in one place, having sex. But was it truly genetic? Ben bred individuals from separate populations, and found that the ability to invade at speed was inherited. Toads evolved as better invaders, in Charles Darwin's words, 'by the accumulation of innumerable slight variations, each good for the individual

---

26   Dubey & Shine 2008.
27   Freeland & Martin 1985.
28   Llewelyn et al. 2010.

possessor'.²⁹ But perhaps this is simply a process of spatial sorting – fitter toads at the front – or perhaps an indication of easier hopping for toads along arterial roads with better drains and maintenance in the Northern Territory?

Whatever the reason, evidence that cane toads are invading faster than ever is a sobering conclusion.

But there is a downside for toads. Big, long-legged toads at the invasion front get severe arthritis of the spine.[30] And in northern Australia, the fastest cane toads ran into the large aggressive native meat ant *Iridomyrmex reburrus*. Native frogs avoid these ants but small cane toads – metamorphs – do not recognise the danger and get eaten alive.[31] By laying catfood bait around waterholes, Rick Shine's Team Bufo amplified the density and killing power of meat ants that killed almost all the metamorphs present.[32] Wrangling meat ants with catfood may not suit the vast savannah of northern Australia, but at least some local fauna are fighting the toxic toad slick.

As it became clear that cane toads were headed as far west as they could go in northern Australia, Professor Shine usefully reviewed evidence of the toad's crimes against ecology (not his words). He concluded that 'no [Australian] native species have gone extinct as a result of toad invasion'.[33] Pleasing as that might be, extinction should not be the benchmark. It is enough to know that the Australian landscape is poorer as a result of the invasion of cane toads.

Rick Shine examined the detail. Where cane toads are abundant they consume large numbers of insects. And cane toads sometimes eat ground-nesting birds like bee-eaters. But large creatures most at risk from the invasion of cane toads are animals with mouths large enough to eat a toad. At risk here are frog-eating frogs and frog-eating snakes (mostly elapid snakes). Death adders, for example, attract prey by wiggling the ends of their tails, it looks like a tasty insect, it attracts cane toads and death adders die when they strike at the toad. Goannas, bluetongue skinks and freshwater crocodiles are all at risk from

---

29 Darwin 2006(1859), p. 741.
30 Cane toad evolution. Cane toads in Oz (n.d.).
31 Ward-Fear et al. 2009.
32 Ward-Fear et al. 2010.
33 Shine 2010, p. 253.

14 Taking the Top End

Figure 14.2 Professor Rick Shine. (Terri Shine. Used with permission.)

eating toads. And northern quolls have become locally extinct when toads arrived. But quolls illustrate the problem of quantifying the cause and effect of cane toads – populations of northern quolls had been in decline long before toads arrived. For Team Bufo, 'separating out the impact of toads from the impact of other threatening processes is a Herculean task in these complex and dynamic systems'.[34]

It's a Herculean task for scientists – of little consequence for an alarmed public. Public alarm has stimulated the distribution of close to $10 million in Australian federal and state government grants and has galvanised enthusiastic locals who contribute time and effort to toadmustering and toadbusting.[35] The visceral response drives people to action. They object to roads being covered with squashed toads; to toads on golfing greens, in swimming pools, playpens and pet food bowls, to the death of faithful dogs, to robbed beehives, to the absence of feisty quolls, to bloated corpses of freshwater crocodiles floating downstream, even to dead snakes, to no more stately goannas stalking the bush, to the loss of native sugarbag honey and other bush tucker.

---

34   Shine 2010, p. 272.
35   Shine 2010, p. 272.

And that is why governments are stimulated to throw money at the most visually prominent 'solutions'.

Government bureaucracies act slowly. Weighty matters to do with threats to the Australian environment come under the Environmental Protection and Biodiversity Conservation Act (1999). For the purposes of the Act, in 2005 the impact of the cane toad on the Australian environment was assessed by the Threatened Species Scientific Committee (TSSC). As cane toads invaded Darwin, the TSSC recommended to the Minister for the Environment and Heritage that they maintain a watching brief.[36] It was likely done in a darkened room because, four years later in October 2009, it was obvious that the cane toad would conquer the entire tropical north of Australia. CSIRO's biological control solutions (e.g. viruses for toads) had not come up with anything promising, for which we should be thankful. At the local level, physical barriers and toadbusting had worked only to protect small islands of fauna. With no real progress in stopping the spreading toad slick, the Australian Government took the lead to identify and protect key habitats, and preserve unique assemblies of biodiversity. In doing so they abandoned the rest of Australia to the march of cane toads.[37] The Australian Government officially threw up its hands and admitted defeat in stopping the spread of cane toads. Peter Garrett, one time vocalist for the environment, then Minister for Environment, Heritage and the Arts, approved a threat abatement plan – a TAP – to provide 'authoritative national leadership' in matters of the cane toad.

This leadership displayed no authority, raised no flag, rallied no followers, and came 74 years too late for the Australian environment.

It is curious to note that in Australia the cane toad has not been declared a pest by either Commonwealth or state governments, the reason being that, if it were declared a pest, landowners would be responsible for control and eradication – an impossible burden.

Where else will cane toads end up? How far will the toad slick spread? In 2008 Mike Kearney and colleagues predicted that cane toads will inhabit around two million hectares of Australia, a semi-circular

---

36   *Threatened Species Scientific Committee* (TSSC) 2005.
37   Minister's decision to develop a threat abatement plan for cane toads. Department of Sustainability, Environment, Water, Population and Communities (n.d.).

area around the north of the continent.[38] They predicted that cane toads will not inhabit the island state of Tasmania. Neither will they inhabit Sydney, Melbourne, Adelaide or Perth. And the nation's chilly capital and seat of government, Canberra, should remain free of cane toads. But these predictions did not last long, and cane toads are now resident in Sydney. Carried to cities inadvertently in freight, toads will most likely find comfortable warm moist spots in the modified climates of urban Australia. Just as, with the help of the cattle industry, cane toads were able to conquer Australia's outback, they will appear in our urban sprawl in the most unexpected places.

It seems that Australian human and native fauna populations have reluctantly accepted cane toads as new migrants, mostly ignored them, sometimes come to blows, sometimes come off the worse for wear, but mostly left each other alone provided they don't interfere in sacred matters like food, drink, sleep, a man's dog, his ute (pickup), fishing and football – and cane toads can interfere with all of these.

Toads, proven survivors, hardened invaders, are evolving and adapting to Australian conditions more rapidly than anyone thought possible. And Australia's naïve fauna is learning quickly. Both invader and invaded are adapting. While this is good news, Professor Shine gives a warning that these rapid changes, happening as we look on, make it very difficult if not impossible to assess the final impact of cane toads, or to predict where they will end up. Understanding the dynamic nature of toads' rapid adaptation provides one of the last hopes for a strategy to protect and conserve remaining habitats unsullied by toads.

There will be many postscripts to this story of invasion. The Lake Eyre basin is one. Lake Eyre, in the arid heart of Australia, is surrounded by deserts and sand dunes. The lake is usually a saltpan lying 15 metres (49 feet) below sea level. When full, it has a surface area around half the size of Lake Ontario. In western Queensland a shallow watershed divides the creeks that flow north into the Gulf of Carpentaria from the creeks that flow south into Lake Eyre. The lake fills with water when rains to the north send floodwaters hundreds of kilometres south. This happens rarely but it happened in 2009, again in 2010 and was topped up once more in 2011. The effect on wildlife was spectacu-

---

38 Kearney et al. 2008.

lar. But the floodwaters carried cane toads. They gobbled up any small creatures they came across and goannas, quolls and snakes died when they tried to eat toads. Angus Emmott of Noonbah Station, just south of Longreach, witnessed the arrival of cane toads in May 2010: 'There's cane toads everywhere. Just to see an invasive animal like this coming in and knocking things about, it's just really sad.'[39]

From Longreach to Lake Eyre the floodwaters made a journey of many months and almost one thousand kilometres as they slowly spread out through the expanse of the inland channel country. These waters carried cane toads into landscapes that nobody ever envisaged they would enter, landscapes that do not appear in predictions of future toad habitats but environments that cane toads can adapt to using stress hormones.[40] As drought follows flood, channels will stop flowing, Lake Eyre will revert back to a saltpan, toads will die in great numbers, but somewhere along the drainage systems some toads will survive and will be ready to spawn when the floods come again – as they surely will.

In 1935, Walter Froggatt prophesied that cane toads would 'reach the river banks and swamp lands of the interior'.[41] And now cane toads can be heard calling in the Kimberley Ranges of Western Australia and the catchment of Lake Eyre in the continent's arid heart. In 1935, few thought this remotely possible.

In the span of three more human generations there is no telling where toads will be, what super-toads will be spawned in the 'Olympic village', what habitats they will invade and what will be the cost for naive native fauna in remote, once pristine ecosystems.

---

39   Cane toad threat to Lake Eyre Basin 2010.
40   Stress helps cane toads' desert invasion 2013.
41   Froggatt 1936, p. 9.

15
Bad, flawed and reckless

Fifteen million years ago a toad died and was entombed on the floodplain of the Magdalena River, northern Colombia, South America – fossil 41159. In the early 21st-century, in a time of rational thought, a time of reason, with mankind spread to the extremities of the planet, a toad lived, anonymous among peers. It squatted, motionless, by a tidal billabong in the broad savannah of the Victoria River, Northern Territory, Australia. Hind legs folded, emaciated brown torso propped on skinny forelegs, hands turned inwards, broad beaked head drooped. Only eyelids moved, slowly blinking each dull eye.

Smoke from a savannah fire layered a cloudless sky, bronzed by the dropping sun. Without a ripple, a saltwater crocodile's triangular snout parted the billabong, evaluated prey, and departed. In the dense canopy of short mangroves, corellas squabbled over position, confettied white petals onto mudcrabs foraging below. No rain for weeks, baked earth desiccating the close of day.

A flooding tide drove steadily inshore, re-wetted tangled mangroves, widened the estuary, re-filled billabongs upstream. This toad, of South American lineage, sired from countless generations, starved, weak, infested by lungworm, caught by the rapid tide, drowned. Mudcrabs dismembered the corpse and made quick work of the feral toad.

This toad died, stranded on the north coast of Australia. Together with its siblings it fled its birthplace. Westwards lay unoccupied territory and toads that went west extended the invasion front. Those that went south dehydrated, dying in arid landscapes. Those that went east ran into competition from older toads and depleted food sources. To the north lay the coast, the barrier of the ocean and death.

Toads stranded on Suriname mudflats gave the toad its name, *Bufo marinus*, the marine appendage ordained by Dutch sailors; observations relayed to the apothecary, Albertus Seba. He called it *Rana marina maxima*, and Carolus Linnaeus, similarly, *Rana marina*. But they were likely on the mudflats out of desperation, not preference. *Bufo marinus* and recently *Rhinella marina* are as much misnomers as cane toad. These toads are no more at home in cane fields than on coastal mudflats. *Rhinella maxima* – giant beaked toad – is perhaps a more apt descriptor. Cane toad is now its universal appellation, not because of habitat preference, but rather because of the folly of the sugar industry. And with the help of man, both the name and the toads are here to stay.

Little more than 80 years after their release into cane fields on the east coast of Australia, despite waterless landscapes, lungworms, predatory goannas, snakes, meat ants and occasional toadbusting humans, cane toads had not far to go to colonise the north of Australia. The invasive toad slick – toxic amphibians travelling in relays – is covering two million hectares of territory.

Many ask who is to blame? Who caused Australia's spreading toad slick? But where do we start? Does blame lie with those who established the sugar cane industry around the world; with the Portuguese who introduced sugar cane to the home of cane toads and other nations who then planted it in the Caribbean, or the Americans who helped establish the sugar cane industry on Hawai'i, or Captains Hope and Whish in Queensland, Australia? Or does blame lie with French and British plantation owners who carried toads from the South American mainland to sugar cane estates in the Caribbean? Or perhaps sugar cane scientists are to blame: the USDA who took toads to Puerto Rico, or the HSPA who took them to Hawai'i, or BSES who delivered cane toads as a panacea for Queensland's cane growers? Where to start? It is a pointless exercise – there is no recourse. Energy is better spent learning from mistakes. If the toad slick in Australia is to have any benefit, it must be to teach lessons to prevent a repeat of similar well intentioned but disastrous events.

To learn a lesson from the cane toad, it is important to understand what went wrong in 1935. How did the practice of well-trained professional scientists, then, compare with what we expect of scientists today? And how much were they driven by a hunger for success, influenced by the industry they served, their political masters and the politics of the times? Has science, scientists and human nature changed since 1935?

With evidence assembled and the advantage of hindsight, when cane toads were released onto the island continent of Australia the premise was bad, the science flawed, and the execution reckless.[1]

Bad premise

The premise of biological control, that toads would control cane beetles and their grubs, was as bad as it gets. In 1935, biological control was touted as the best in pest control, more effective than poison, pitch or prayer. But biological control is a flawed concept. Its unspoken premise is that an organism, once released, will understand the altruism of its liberation, that it will show gratitude for the confidence bestowed, that it will understand that it must eat only what we want it to eat, that it will eat no other foods no matter how tempting. It will agree, once the food runs out, that it will eat its offspring and then die of starvation, that it will not wander the landscape sampling other delights. It will understand that both it and its offspring are jointly and severally liable for their behaviour, and that we will seek injunctive relief for any transgression.

We try to control intelligent beings. We can direct and hope to control rational adult humans by economic means or, for a short period, by coercion. We can direct young children, for a short period at best. We can train a dog but must constrain it when we finish. So what hope do we have of controlling insects – or toads?

Biological control is the stuff of fairy stories; the practice biological, the control a fantasy.

One problem with toads is that they are general feeders, eating anything they can swallow, and proponents of biological control might argue that being generalist feeders is their fatal flaw as agents of pest control – their profligate habits should not condemn the entire dis-

---

1   Some of the ideas in this chapter were first published in Turvey 2010.

cipline. Organisms that have only one food source – the target plant or insect – may be better candidates for biological control agents. But when the food source is gone, or novel food sources appear, these organisms can change their dietary preferences. Cane toads are not the only disasters to be laid at the door of biological control.[2]

The bug, *Teleonemia scrupulosa*, introduced to Africa to control *Lantana* weed jumped across to sesame crops in Uganda, devastating farmers. The giant African snail, *Achatina fulica*, introduced to Hawai'i for food and medicinal purposes, became a pest; a species of carnivorous land snail, *Euglandina rosea*, introduced to Hawai'i in the 1950s to control *Achatina*, invaded native forests causing the extinction of the native Oahu tree snail *Achatinella mustelina*. A tachinid fly, *Bessa remota*, introduced to Fiji to control the coconut moth, *Levuana iridescens*, was so effective that the moth, the only member of its genus, became extinct. Dung beetles were introduced to Hawai'i to bury cattle dung and so reduce breeding sites of the horn fly. But when dung beetles populated upland pastures they provided a food source for the mongoose such that it could extend its killing range into mountain habitats. Japanese interests introduced goannas (monitor lizards) to islands in Micronesia to control rats, but when goannas became pests themselves the proponents introduced cane toads to poison them. Now rats, goannas and toads are all pests on these islands. The *Mixoma* virus, introduced to Great Britain to control rabbits, was so effective that rabbit populations declined rapidly. A cascade of environmental effects followed. Pasture habitats for the large blue butterfly, *Maculina arion*, for its symbiotic partner, a species of red ant, *Myrmica sabuleti*, and for the butterfly's food source – wild thyme – also changed rapidly in response to the removal of rabbits. Added to this, pasture management and cattle grazing practices had changed over the years. Combined changes in habitat were so great that in 1979 the large blue butterfly became extinct, regionally – it has since been reintroduced.

But perhaps the most shocking case of biological control gone bad is the postscript to the story of *Cactoblastis cactorum*, the moth that destroyed prickly-pear in Australia. It was one of the most successful biological control agents in history – a miracle moth – up there with

---

2   Howarth 1991.

the vedalia beetle, Albert Koebele's ladybird. Alan Dodd adopted the moth from its homeland in Argentina and, because of the care in selection, in Australia it targeted its host and no others. Prickly-pear was wiped from the landscape and exists in Australia as isolated plants that continue to provide a food source for the moth. There were no native *Opuntia* cacti in Australia, so the release of the moth after testing was both rational and reasonably safe.

Because of its success in Australia, *Cactoblastis* was introduced to control pest cacti in sub-Saharan Africa, New Caledonia, Hawai'i, Mauritius, the Cayman Islands, St Helena, Ascension Island, Pakistan and islands in the Caribbean.[3] From the Caribbean, in 1989, it entered Florida and the North American mainland. The moth is now threatening both indigenous and cultivated *Opuntia* cacti in the United States and Mexico and it could have 'a devastating effect on the landscape and biodiversity of native desert ecosystems'.[4] It could also devastate commercial and subsistence cropping of *Opuntia* in Mexico where these cacti are used as vegetables, fruit, and for making alcoholic drinks. But Tequila drinkers are safe – for the moment. Tequila is made from the succulent *Agave*. Despite artificial pheromones – sexual attractants – to lure moths into traps, and release of sterile males who sire no offspring, these moths cannot be eradicated and are certain to have ecological and economic consequences.

One day a miracle moth – the next day a pest. And so it goes for biological control.

Today, there is less talk about biological control and more talk about integrated pest management, or IPM in current parlance. IPM is a broad church built around maintaining economic crop yields first and foremost. Local organisms are encouraged to establish themselves in and around the crop so that they can control outbreaks of pests. And if it all gets too much for the locals, then there are always chemicals that can be targeted in minimal amounts to the right pest at the right time and not result in a crop growing in a biological desert. It is an intelligent application of both sophisticated agricultural chemicals and the best features of biological control.

---

3   Zimmermann et al. 2000.
4   Zimmermann et al. 2004.

Flawed science

Raquel Dexter, in her paper presented to the Puerto Rico conference in 1932, stated that toads controlled populations of May beetles. But her conclusion exceeded the bounds of her experiment – it was flawed. She showed only that toads ate beetles – and pretty well anything else small enough to swallow. Without knowledge of natural variation in beetle populations caused by fecundity, disease, predation, soil condition and seasons, she could not conclude that toads controlled populations of beetles. There was a logical chasm between cause (toads found to eat beetles) and effect (reduced populations of soil-dwelling beetle larvae). She concluded, wrongly, that because toads had eaten beetles on the day they were collected, they were responsible for controlling both larvae and adult beetle populations throughout the year. Raquel Dexter imitated Archie Kirkland's survey of the contents of toads' guts.[5] But this USDA scientist made only generalised conclusions about the usefulness of toads to farmers; on balance they ate more injurious creatures than beneficial ones – the reason why French and British farmers also sought their services. He did not confer toads with the ability to control populations of garden pests. Dexter's conclusion was a basic error – flawed science.

Moreover, this flaw was not flagged by her colleagues. Cornell-trained Wolcott, Stanford-trained Pemberton and Harvard-trained Bell all believed Chicago-trained Dexter's experiment showed that cane toads controlled populations of May beetles in cane fields. Dexter's paper, included in the Proceedings of the Congress, established the myth of the toad's scientific pedigree. And the myth went unchallenged. At this professional level, it was a failing of both science and scientific inquiry.

It is an inexplicable conundrum, hard to understand, why none of the professional scientists who either heard or examined Dexter's paper, neither scientists attending the Puerto Rico Congress, nor in HSPA, nor BSES, exposed the flaws of logic between observations and conclusions. None exposed the flawed science. Instead, Dexter's paper on the feeding habits of *Bufo marinus* exerted a remarkable influence far

---

5   Kirkland 1904.

exceeding its merit. It helped convince Cyril Pemberton of the benefits of importing toads to Hawai'i, it impressed Arthur Bell and Reg Mungomery, and it was touted by BSES in defence of the toad to the Australian Commonwealth Department of Health. This innocuous paper, reporting imitative research, delivered by a little known researcher at a concurrent conference session, was endowed with influence far beyond its subject matter – a survey of the stomach contents of 301 toads.

BSES's own entomologists James Illingworth and Alan Dodd had described the swarming habits of cane beetles. The observations from Alan Dodd's solitary night-time vigils in Queensland's cane fields, alone with his demons from the Western Front, were recorded in their landmark 1921 monograph.[6] Both entomologists left BSES soon after. In September 1933, Bell asked his entomologists their opinion of the toad's suitability for controlling canegrubs but, just a step away from his office, in the library of the Department of Agriculture and Stock in Brisbane, Bell had the answer to his question. Even armies of toads, had they been inclined to wait in cane fields for the short nuptial swarming of cane beetles, could never have reduced beetle populations. Toads could have had no impact at all on cane grubs, the beetles' soil-dwelling offspring and pest of sugar cane crops. If they had read their own organisation's research bulletins, Bell and his colleagues would have understood the impossible task faced by toads in controlling populations of cane beetles and grubs. Reviewing published research is one of the building blocks of scientific endeavour, drummed into researchers at an early age, so there is little defence for this crime of omission.

Scientists in the USDA, HSPA and BSES promoted toads. But scientists are not a homogenous breed. The science philosopher Sir Peter Medawar described them well: 'Among scientists are collectors, classifiers, and compulsive tidiers-up; many are detectives by temperament and many are explorers; some are artists and others artisans'. And scientists tell stories and publish them, 'stories which might be about real life, but which have to be tested very scrupulously to find out if indeed they are so'.[7] Among these are the breed of applied scientists who populated the USDA, HSPA and BSES. Their brand of applied science is

---

6 Illingworth & Dodd 1921.
7 Medawar 1969, pp. 133, 142, 148, 169.

one of the most rewarding because it delivers benefits locally. Improved cane yields linked to a more profitable industry and more local employment is evident reward for research. Benefitting industry is the job of these scientists. Some will argue that, in the case of the toad, this breed of applied scientist did not follow the 'scientific method' of hypothesis and testing. But that is how science is *reported*, not necessarily how it is *done*. Science is rather more haphazard and the wastelands of scientific endeavour are dotted with blooms of serendipity. In the Top End of Australia, the ecological groundwork done at the Daly River and Fogg Dam prior to the invasion of cane toads was crucial to quantifying post-invasion impacts, but the genesis of each program lay elsewhere – the modern conclusions serendipitous – but good science nevertheless. But in the toad's case, there is no excuse for complete absence of testing the organism before release into its new environment in 1935.

Certainly, the science behind the introduction of the toad was poor by today's standards, but the number of well-qualified scientists supporting the toad in 1935 indicates that the standard of science was acceptable for the times. And, as the Manhattan Project that delivered the atomic bomb demonstrated, ten years later, good science will deliver a good scientific outcome, but it may be neither benign nor humanitarian.

Sugar cane scientists – applied scientists – embedded within the industry they serve, can be divorced from neither the politics of their industry nor their times.

At the Insular Experiment Station at Rio Piedras, scientists tried everything they knew but failed to satisfy demands from the sugar industry and defeat white grubs of May beetles. Their future employment was bleak. Toads, when they arrived with their apparent powers to control white grubs, became saviours of both the industry and the scientists. Perhaps scientists wanted a saviour so badly that, once found, they did not scrutinise their saviour's powers too closely.

In Hawai'i, Cyril Pemberton, world-renowned entomologist and applied scientist, was clearly under pressure from John Waldron, both his boss in HSPA and a leading figure in the sugar industry. Cautious Pemberton took his time to review the toad in Puerto Rico but once it was released in Hawai'i, although it did not control anomala beetles, Pemberton helped distribute it around the Pacific, supported its introduction to Australia, and mounted a bufo wind vane on his house –

a testament to his enthusiasm for the anuran. But for this careful scientist, the release of an otherwise untested organism, together with a hidden lungworm parasite, was a tectonic fracture through the continent of his craft. The lesson: well respected, successful scientists are also fallible. And, like all of us, scientists are ephemeral.

In Honolulu, HSPA's head office is now a community centre, its toad-rearing station at Waipio under boulevards of new housing estates. The Hawaiian sugar industry, begun in 1825, has gone and only one sugar mill remains in production.[8] But agents of biological control introduced by the likes of Koebele, Perkins, Muir and Pemberton have persisted far longer than the industry they were supposed to help. Wasps, flies, bugs, mongoose and toads are their legacy to Hawai'i; and cane toads their gift to Australia.

Reckless execution

By recklessly releasing toads in Australia, with no prior testing, relying only on reports from Barbados, Puerto Rico and Hawai'i, Bill Kerr, Arthur Bell and Reg Mungomery ignored the excellent protocols for biological control established by Alan Dodd and the Prickly-pear Board and impressed on a young Mungomery. They also ignored and dismissed Walter Froggatt's clear warnings. But although Froggatt was prescient, he was not correct. Froggatt worried about the consequences of what the toad would eat – insects, ground-nesting birds, lizards, frogs – in this he was proved right. But he did not seem to appreciate the dire consequences for Australia's unique fauna should they eat toxic toads. This was the main problem. In 1935 nobody seemed to understand that some naive Australian native fauna would try to eat toxic cane toads and die – did not understand the consequences of releasing a toxic amphibian into a post-Gondwanan refuge. Mungomery discovered in Honolulu that *Bufo marinus* was poisonous when members of a Philippine family died from eating a toad. He expressed his concerns to Arthur Bell, worried enough to keep it a close secret, worried a little that Australians might eat it by mistake, but worried not at all

---

8   The Hawaiian Commercial and Sugar Company on Maui (n.d.).

about the impact of the toad on Australian fauna that may try to eat it. He forged ahead, forgot Dodd's training, forgot his own words written in 1934 about poorly executed biological control and about 'upsetting of the whole biological balance'.[9] And he forgot the hard lessons learnt in Childers cane fields, that cane beetles were active only briefly after emerging from the soil. He forgot why he rightly dismissed the value of the toad at first. Had he held to his opinion, thoughts of the cane toad would have gone no further – Australia may have been spared the toad. But Mungomery was distracted by the prospect of a trip to Hawai'i and won over by Dexter's paper expounding the toad in Puerto Rico. That is when things turned bad for Australia.

With directives from Bill Kerr and Arthur Bell in BSES, with encouragement from Cyril Pemberton in Hawai'i, with the confident assertions of Dexter's paper and with the Congress of the International Society for Sugar Cane Technologists about to be hosted by BSES in Brisbane within days, Mungomery released toads into the landscape around Gordonvale in north Queensland. The eventual fate of Australian fauna was far worse than what David Rivett, head of Australia's peak science organisation CSIR, called 'the decidedly pessimistic forecast of the New South Wales entomologist'.[10]

But these were not the sole factors. Releasing toads was supported by a lineage of similar farming practices. The 19th century was particularly busy. In France and Britain, farmers traded toads to control pests of market gardens, and Caribbean plantation owners – French in Martinique and British in Barbados – imported giant toads from the South American mainland to control pests of their cane fields. It was also the era of Francis Trevelyan Buckland, co-founder of the Society for the Acclimatisation of Animals, Birds, Fishes, Insects and Vegetables within the United Kingdom,[11] and the Marquis of Breadalbane, farmer of eland, yak and bison in Scotland. Eugene Schieffelin of The American Acclimatisation Society attempted to introduce to North America every bird mentioned in the plays of William Shakespeare, including the pestiferous European starling.[12] And The Queensland Ac-

---

9 Mungomery 1934b.
10 Rivett 1935.
11 Lever 1992, p. 29.
12 Ehrlich et al. 1988.

climatisation Society was responsible for the release in 19th century Queensland of fallow deer, axis deer, red deer, rusa deer, angora goats, llamas, hedge sparrows, skylarks, blackbirds, song thrushes, rooks, house sparrows and starlings, and prickly-pear. Their combined missions were to increase populations of feral animals and plants and they worked hard at it.

Just 73 years before Reg Mungomery released toads at Gordonvale, new colonists were busy clearing tracts of Queensland's coastal forest and attempting to cultivate 'most, if not all, the productions of the Indies, South America, and not a few of those of Africa'.[13] This was coastal land of which Captain Cook, the 18th-century explorer, said 'naturally produces hardly anything fit for Man to eat, and the Natives know nothing of Cultivation'.[14] It was a licence for settlers to clear forests and 'improve' the land. BSES was in the business of improvement as was HSPA in Hawai'i, the USDA at Mayagüez in Puerto Rico and its poor cousin, the Insular Experiment Station at Rio Piedras. Biological control was the latest improvement fashion, and toads the latest fad. When toads were brought to Queensland, it was unremarkable. It was common practice for sugar cane growers.

In Australia, it was Arthur Bell's idea and his decision to import *Bufo marinus*. In spring of 1933 with rain falling and cane grubs soon to pupate, Bell put himself under pressure to help his clients, Queensland's cane growers. Importing the toad was covered by Cumpston's quarantine permit. There was neither breach of, nor failure to follow, quarantine procedures at the time. There was no need for political manoeuvring. Politics came into play only when Cumpston, persuaded by Froggatt and his allies, changed his mind and banned the toad. Kerr determined to get Cumpston's ban lifted, called in the heavy artillery of the sugar industry and mounted a punishing assault on Cumpston to complete the introduction of cane toads to Queensland.

Between 1932 and 1935, there was a great deal of political and economic pressure both in Australia and internationally. The world's sugar technologists met in Brisbane under what its president called 'a continuance of the acute economic depression which has enveloped the entire world'.[15] President Franklin D. Roosevelt was implementing his

---

13 Pugh 1863.
14 Wharton (ed.) 2012.

New Deal to lift America out of economic depression, and Adolf Hitler was rearming Germany, flaunting the Versailles Treaty and stirring up talk of war. A mind-set of war prevailed in Australia. Sugar regulations gave financial encouragement for white labour to populate tropical northern Australia with 'a defensive garrison' against a 'temptation to Asiatic invasion',[16] and BSES scientists believed they were engaged on behalf of mankind in 'continuous warfare upon the insects' that attacked the cane.[17]

Global economic depression and impending war was the milieu of Cyril Pemberton and John Waldron of HSPA, of Bill Kerr, Arthur Bell and Reg Mungomery of BSES, Bill Doherty of Queensland Cane Growers' Council, Robert Veitch who was Queensland's Chief Entomologist, David Rivett the Head of CSIR, John Cumpston the Director General of Health, Arthur Graham the Queensland Under Secretary for Agriculture and Stock, Frank Bulcock the Queensland Minister for Agriculture and Stock, William Forgan Smith the Premier of Queensland, and Joseph Lyons the Prime Minister of Australia. They were men of their times, on a mission to help the sugar industry and protect the economy that depended on it, not evil-doers bent on populating Australia with a toxic pest. And that makes a repeat of their actions all the more difficult to prevent.

It is far easier to police the expected mayhem of madmen than to guard against the misguided actions of men of high standing.

The cane toad was championed by well-trained, reputable scientists committed to helping their clients in the sugar industry, assisted by inadequate quarantine regulations and supported by politicians who relied on the strength of the sugar industry for their continued existence. Politics was too closely woven into the fabric of the sugar industry to allow objectivity to illuminate science. For some, the promise of the toad was enough. Hope clouded judgement, halted closer scrutiny, permitted poor science. At the time, there was little understanding of ecological consequences – ecology and biogeography were poorly understood – but there were critics enough. Sharp, and for a time Perkins in Hawai'i and Froggatt in Australia were critical of biological control,

---

15 Gibson 1935, p. 35.
16 Anon. 1936.
17 Kerr & Bell 1939, p. 172.

of the impact of the release of untested organisms, and they likely had many allies.

Could it happen again?

Could we have a modern equivalent of the cane toad? Could a destructive alien organism championed by respected scientists, government departments and politicians, be released today? There are clear signs that it could.

One mechanism for release is through enthusiastic meddling with animal populations. Ideas like Pleistocene Rewilding, 'a bold plan ... that aims to restore some of the evolutionary and ecological potential that was lost 13,000 years ago'.[18] Pleistocene Rewilding refers to the Pleistocene geological epoch, the first part of the Quaternary geological period that lasted from around 2.6 million years ago to 12,000 years ago and encompassed recent glaciations in the northern Hemisphere. During the Pleistocene, large mammals like mammoths, Irish elk, sabre toothed tigers and cave bears became extinct. The objective of Pleistocene Rewilding is to redress this loss, to restore large mammals to the landscape – to introduce similar-sized creatures with similar habits to those lost during the Pleistocene. To release Asian asses, Przewalski's horse, Bactrian camels, Asian and African elephants, cheetahs and lions to roam free in public and private spaces in North America, in preference to 'rats and dandelions that will otherwise come to dominate the landscape'.[19]

Promoters of Pleistocene Rewilding at Cornell University, Josh Donlan and colleagues, admit that the obstacles are substantial and the risks are not trivial. Pleistocene Rewilding will make for interesting times. Cattlemen and ranchers complain about reintroduction of wolves into wild areas of the United States, so imagine the outcry ensuing from release of a surrogate for a sabre toothed tiger!

The fingerprints of 19th-century acclimatisation societies are clearly visible.

---

18 The Rewilding Institute (n.d.).
19 Donlan et al. 2005.

Today, Australia's borders are protected by strict quarantine.[20] No-one could legally import toads into Australia the way Reg Mungomery did. The premises of origin and the receiving premises must be managed by approved, licensed or registered organisations, and the organism, together with its habitation and its waste, must be isolated and destroyed once used. But novel organisms created by genetic manipulation – genetically modified organisms or GMOs – are a different matter. The anticipated benefits of GMOs to increase crop yields using less land, water, energy and agricultural chemicals, are good things. They can improve food security, address rural poverty and benefit both man and environment. But, as with biological control, the problem is in the *unexpected* rather than the *expected* consequences of the release. In Australia, GMOs and their release are regulated under the *Gene Technology Act (2000)* through the Office of the Gene Technology Regulator (OGTR). But regulation of GMOs is based on managing risk, on definitions of thresholds of acceptable risk. Release is not based on absolutes – creating no harm – but on probabilities – no likelihood of creating harm.

In defining acceptable risk, the OGTR is advised by three committees on scientific, technical, ethical, and community matters.[21] Likewise, in 1935 both state and Commonwealth governments were advised by the best available scientists. Among these scientists, Walter Froggatt's predictions were considered 'decidedly pessimistic' by Sir David Rivett, head of CSIR, Australia's premier science body, and 'radical and biologically impossible apprehensions' by Cyril Pemberton, Chief Entomologist of the world-renowned HSPA. And Queensland's Chief Entomologist, Robert Veitch, supported the toad and the decision of the Director of BSES, Bill Kerr. In 1935, advice from leading scientists to a body like the OGTR, had it been in place, would most likely have been to allow the toad's release because of its apparent low risk.

From here, it is no great leap of imagination to think of a GMO, evaluated as low risk, promoted by industry and government, and released by enthusiastic and myopic scientists, turning into a mega-weed, resistant to weedicides, growing vigorously on saline soils, coping with

---

20 Import Conditions, Department of Agriculture, Fisheries and Forestry (n.d.).
21 Australian Government Department of Health and Ageing. Office of the Gene Technology Regulator (n.d.).

drought stress, reproducing abundant and viable seed at will and taking over landscapes – all traits that are candidates for genetic transformation.

Synthetic biology – synthesising new organisms from novel genetic code – is even more worrisome. All that is needed to create a novel organism is a DNA sequencer, a desktop computer and a 3D printer with its own 'inkjet' proteins. No need to carry organisms around the world in a suitcase; just download the genetic code from the other side of the world and synthesise your own. Compared to this technology, creation of GMOs by gene splicing and cloning seems like a huffing and puffing steam engine.

But even without new technologies, as demonstrated by the cane toad, a novel organism could once again be released into an unsuspecting environment simply by convincing the gatekeepers of science. Just consider BSES's 1935 proposal in the light of modern criteria for research funding and imagine the PowerPoint presentation in front of funding agency bureaucrats or well-heeled investment angels:

The cane toad

- builds on successes in biological control
- replaces toxic and residual pesticides
- is supported by a published scientific paper
- has international scientific peer review
- is endorsed by Australia's peak science body
- is championed by industry
- is promoted by the Queensland Government and its Premier
- is approved for use by the Commonwealth Government and
- has personal endorsement from the Prime Minister.

It is a short odds winner – straight to the top of the funding list – yes, it could happen again.

Today we have the best scientists on the job, but the best scientists were also at work in 1935. It is manifest hubris and simply wrong to think that we are qualitatively different. In 80 years time or more, our descendants will likely be appalled by our profligate use of energy and water, the gross imbalance of our carbon economy, the naivety of our genetic manipulation and still wondering why cane toads are everywhere.

The toad was championed by a herd of supporters, scientists, industry people and governments. But in each herd there are people like Walter Froggatt who can see further than the rest, who can see the precipice, but whose voice is lost in the thunder of hooves. In a stampede the collective ego is blind, the herd instinct is contagious and naysayers get trampled. In the 1930s, the herd was spooked by economic depression, the threat of Asian invasion and the need to defend the canefields and the north – cane toads were recruited as fellow travellers. Today, cane toad surrogates are present in debates on climate change, GMOs, complementary medicine, vaccination, and a host of other causes. The following herds are unnerved by contrary ideas. The lesson, then, is not to join the herd but graze greener pastures, encourage difference and let ideas either flourish or wilt under scrutiny. This much we can learn from the cane toad.

Through no fault of its own, *Bufo marinus*, *Rhinella marina*, cane toad, failed in its allotted task to control cane grubs; failed in the Caribbean, in Hawai'i and in Queensland. Politicians, cane farmers and scientists alike saddled this primitive amphibian, toxic coloniser, with their own aspirations. They carried it to new territories where they left it to fend for itself, to come to terms with its new environment, to flee competition from siblings, to forage in new habitats. Toads did what they knew best: radiated out, occupied new ground, ate what was available and bred frenetically. With the help of man, the cane toad is now among the world's top 100 invasive species.[22]

Their release in Australia was bad, flawed and reckless, a failing of cane-farmers and farming, of government and governance and a failing of science and scientists. Exotic toads have been part of the Australian landscape for three human generations and look like they are here to stay – toxic, unlovable and immovable.

---

22 Listed as *Rhinella marina* in: Global Invasive Species Database (n.d).

Works cited

Primary sources

Anon 1935. Proceedings of the fifth congress of the International Society of Sugar Cane Technologists, Brisbane, 27 August to 3 September, p. 406.
Archer M 2013. Professor of PalaeoBiology, University of New South Wales. Personal communication, 3 June.
Assistant Director BSES[1] 1936a. Letter to RW Mungomery, 21 September. BSES File 507-0000. Distribution. 1935–1950.
Assistant Director BSES 1936b. Letter to the Director General of Health, 31 August. BSES File 507-0000. Distribution. 1935–1950.
Assistant Director BSES 1936c. Letter to WW Froggatt, 31 March. BSES File 507-0000. Distribution. 1935–1950.
Assistant Director BSES 1937a. Memorandum to RW Mungomery, 5 January. BSES File 507-0000. Introduction. 1933–1941.
Assistant Director BSES 1937b. Memorandum to RW Mungomery, 28 May. BSES File 507-0000. Toads General. 1935–1949.
Assistant Director BSES 1938. Memorandum to Undersecretary, 19 January. BSES File 507-000. Toads General. 1935–1949.

---

1   BSES is an abbreviation of Bureau of Sugar Experiment Stations, a department of the former Queensland Government Department of Agriculture and Stock.

Assistant Director BSES 1939. Letter to RW Mungomery, 19 December. BSES File 507-000. Toads General 1935–1949.

Assistant Entomologist BSES 1936. Letter with attachments to AF Bell, 26 March. BSES File 507-0000. Distribution. 1935–1950.

Bell AF 1917. World War I War Service Record, National Archives of Australia. Online: naa12.naa.gov.au/scripts/imagine.asp?B=3008335&I=1&SE=1. Retrieved 7 August 2013.

Bell AF 1928. Typed report on traveling scholarship, dated 22 March 1928. Personnel File, Arthur Bell, BSES.

Bell AF 1936. Memorandum to RW Mungomery, 21 September. BSES File 507-0000. Distribution. 1935–1950.

Bell AF 1937. Telegram to RW Mungomery, 11 December. BSES File 507-000. Toads General. 1935–1949.

Bell AF 1938. Memorandum to RW Mungomery, 1 February. BSES File 507-000. Toads General. 1935–1949.

BSES 1940. Annual Report, Queensland Department of Agriculture and Stock, Bureau of Sugar Experiment Stations. Brisbane.

Byrne T 1988. Interviewed in Lewis M 1988. Stamp's Road Lagoon, Tully, Queensland. 18 February.

Chief of the Division of Entomology, CSIRO 1972. Letter to Director, Bureau of Sugar Experiment Stations, 12 May. BSES File 507-0000. Toads Distribution. 1961–1981.

Cilento RW 1935. Certain social oddities in North Queensland. Proceedings of the fifth congress of the International Society of Sugar Cane Technologists, Brisbane. pp. 28–34.

Cumpston JHL 1935a. Letter to Undersecretary, Queensland Department of Agriculture and Stock, 4 December. BSES File 507-0000. Distribution. 1935–1950.

Cumpston JHL 1935b. Telegram to Undersecretary, Queensland Department of Agriculture and Stock, 4 December. BSES File 507-0000. Distribution. 1935–1950.

Dexter RR 1929a. Letter to FR Lillie, 8 July. Frank R Lillie Papers 1910–1931. Box 3 Folder 3. Special Collections Research Centre, University of Chicago Library.

Dexter RR 1932. The food habits of the imported toad *Bufo marinus* in the sugar cane sections of Porto Rico. Proceedings of the fourth congress of International Sugar Cane Technologists, San Juan, Puerto Rico, Bulletin 74. pp. 1–6

Director BSES 1934. Memorandum to Undersecretary, Department of Agriculture and Stock, 11 December. RW Mungomery Personnel File, BSES.

Works cited

Director BSES 1935a. Letter to Director General of Health, 25 March 1935. BSES File 507-0000. Introduction. 1933–1941.

Director BSES 1935b. Letter to D Rivett, 5 December. BSES File 507-0000. Distribution. 1935–1950.

Director BSES 1935c. Letter to FCP Curlewis, 4 December. BSES File 507-0000. Distribution. 1935–1950.

Director BSES 1935d. Memorandum to RW Mungomery, 12 November. BSES File 507-0000. Distribution. 1935–1950.

Director BSES 1935e. Memorandum to Undersecretary, Department of Agriculture and Stock, Brisbane, 15 March. RW Mungomery Personnel File, BSES.

Director BSES 1935f. Memorandum to Undersecretary, Department of Agriculture and Stock, 9 November. BSES File 507-0000. Distribution. 1935–1950.

Director BSES 1935g. Memorandum to Undersecretary, Department of Agriculture and Stock, 11 November. BSES File 507-0000. Distribution. 1935–1950.

Director BSES 1937a. Memorandum to RW Mungomery, 21 April. BSES Files 507-000. Toads General. 1935–1949.

Director BSES 1937b. Memorandum to Undersecretary Department of Agriculture and Stock, 13 December. BSES File 507-000. Toads General. 1935–1949.

Director BSES 1972. Letter and appended note to the Director General, Department of Primary Industries, 24 April. BSES File 507-0000. Toads Distribution. 1961–1981.

Dodd AP 1916. World War I War Service Record, National Archives of Australia. Online: naa12.naa.gov.au/scripts/imagine.asp?B=3510167&I=1&SE=1. Retrieved 15 August 2013.

Dodd AP 1926. Letter to the Secretary, Commonwealth Prickly-pear Board, 12 January. National Archives of Australia. Prickly Pear investigations – Cactoblastis opuntia aurantiaca 1924–1926. Series No A8510. Control 81/85. Barcode 1933359.

Dodd AP 1945(1917)a. War Diary. 16 June. Typed transcription of hand written diaries of Alan Dodd. Queensland Museum Division of Entomology, Brisbane.

Dodd AP 1945(1917)b. War Diary. 18 October. Typed transcription of hand written diaries of Alan Dodd. Queensland Museum Division of Entomology, Brisbane.

Dodd AP 1945(1918). War Diary. 27 July. Typed transcription of hand written diaries of Alan Dodd. Queensland Museum Division of Entomology, Brisbane.

Doherty WH 1935. Letter to W Kerr, 25 November. BSES File 507-0000. Distribution. 1935–1950.
Eckford S 2012. President, Julia Creek Historical Society. Personal communication, 4 December.
Freeland WJ 1984 *Cane toads: a review of their biology and impact in Australia*. NT Parks and Wildlife Unit, Conservation Commission of the Northern Territory, Winnellie.
Froggatt WW 1936a. Letter to AF Bell, c. April. BSES File 507-0000. Distribution. 1935–1950.
Gibson AJ 1935. Presidential Address. Proceedings of the fifth congress of International Sugar Cane Technologists, Brisbane.
Goldfinch PHM 1935. Economics of sugar in Australia. Proceedings of the fifth congress of the International Society of Sugar Cane Technologists, Brisbane.
Harold L Lyon Arboretum. University of Hawai'i (n.d.). Online: www.hawaii.edu/lyonarboretum/. Retrieved 6 September 2013.
Hawaiian Sugar Planters' Association 1933. Monthly reports of the Entomology Department for August, September, October, November, December. Bound typescripts, HARC-HSPA.
Hawaiian Sugar Planters' Association 1934a. Monthly report of the Entomology Department for February. Bound typescripts, HARC-HSPA.
Hawaiian Sugar Planters' Association 1934b. Report of Committee in charge of the Experiment Station Hawaiian Sugar Planters' Association for the year ending 30 September.
Hawaiian Sugar Planters' Association 1935a. Monthly report of the Entomology Department for June. Bound typescripts, HARC-HSPA.
Hawaiian Sugar Planters' Association 1935b. Monthly report of the Entomology Department for May. Bound typescripts, HARC-HSPA.
Hawaiian Sugar Planters' Association 1935c. Monthly report of the Entomology Department for November. Bound typescripts, HARC-HSPA.
Hawaiian Sugar Planters' Association 1936. Report of Committee in charge of the Experiment Station Hawaiian Sugar Planters' Association for the year ending 30 September.
Hawaiian Sugar Planters' Association 1937. Monthly report of the Entomology Department for September. Bound typescripts, HARC-HSPA.
Hawaiian Sugar Planters' Association Experiment Station 1920. Director's Report, October, November, December.
Hawaiian Sugar Planters' Association Experiment Station 1921. Director's Report, June.
Holroyd P 2009. Museum of Paleontology, University of California, Berkeley. Personal communication.

Works cited

Hope L. 1844. Journals. In the possession of Mrs P Hiley, Byways Steep, Petersfield, Hampshire. Australian Joint Copying Project. National Library of Australia. State Library of New South Wales. 1978. Microfilm.

Illingworth JF & Dodd AP 1921. Australian sugar cane beetles and their allies. Division of Entomology, Bulletin 16. Bureau of Sugar Experiment Stations, Brisbane.

Ingram JW, Jaynes HA & Lobdell RN 1938. Sugarcane pests in Florida. Proceedings of the sixth congress of the International Society of Sugar Cane Technologists. Louisiana State University, Baton Rouge, Louisiana, 24 October to 5 November.

Jarvis E 1934. Notes on the Toad Bufo marinus L. Typewritten report. BSES File 507-0000. Introduction. 1933–1941.

Jepson WF & Moutia LA 1938. The progress of applied entomology in Mauritius during the years 1933 to 1938, with reference to insects of the sugar cane. Proceedings of the sixth congress of the International Society of Sugar Cane Technologists, Louisiana.

Kerr HW 1931. Letter to HT Easterby, 26 September, Bureau of Sugar Experiment Stations. Arthur Bell, Personnel File.

Kerr HW 1934. Thirty-fourth Annual Report of the Bureau of Sugar Experiment Stations. 14 December.

Kerr HW 1935a. Letter to Australian Sugar Producers' Association, Queensland Cane Growers Council, 23 November. BSES File 507-0000. Distribution. 1935–1950.

Kerr HW 1935b. Thirty-fifth Annual Report of the Bureau of Sugar Experiment Stations. 15 November.

Kirkland AH 1904. Usefulness of the American Toad. *Farmers' Bulletin*, 196. US Department of Agriculture, Washington.

Koebele A 1890. Report of a trip to Australia to investigate the natural enemies of the fluted scale. US Department of Agriculture, Division of Entomology, Bulletin 21.

Kurth AA 1942. Letter to Undersecretary, 26 April. BSES File 507-000. Toads General. 1935–1949.

Lee AK & Straughan IR 1961. Report to the United Graziers Association on a survey of stock losses attributed to the Cane Toad (*Bufo marinus*). Unpublished Report, Zoology Department, University of Queensland. BSES File 507-0000. Toads General. 1961–1981.

Lee HA 1927. Pathology Department monthly report, August. Hawaiian Sugar Planters' Association Experiment Station, 10 September 1927.

McDougall W 1938. Memorandum to Assistant Director, BSES, 27 January. BSES File 507-0000. Introduction. 1933–1941.

Mungomery RW 1933. Letter to AF Bell, 28 September. BSES File 507-0000. Introduction. 1933–1941.
Mungomery RW 1934a. Memorandum to Kerr HW, 15 November. BSES File 507-0000. Distribution. 1935–1950.
Mungomery RW 1935a. A short note on the breeding of *Bufo marinus* in captivity. Proceedings of the fifth congress of the International Society of Sugar Cane Technologists, Brisbane. pp. 589–91.
Mungomery RW 1935b. Letter to AF Bell, 10 January. BSES File 507-0000. Introduction. 1933–1941.
Mungomery RW 1935c. Letter to AF Bell, 28 April. Pleasanton Hotel, Honolulu. BSES File 507-0000. Introduction. 1933–1941.
Mungomery RW 1935d. Memorandum to AF Bell, 1 July. BSES File 507-0000. Toads General. 1935–1949.
Mungomery RW 1936. Memorandum to Dr H Kerr, 30 December. BSES File 507-0000. Introduction. 1933–1941.
Mungomery RW 1937. Telegram to AF Bell, 11 December. BSES File 507-000. Toads General. 1935–1949.
Mungomery RW 1938. Memorandum to AF Bell, 11 January. BSES File 507-000. Toads General. 1935–1949.
Mungomery RW 1939. Memorandum to AF Bell, 28 March. BSES Files 507-000. Toads General. 1935–1949.
Pathologist BSES 1933. Memorandum to WA McDougall, E Jarvis & RW Mungomery. 26 September. BSES File 507-0000. Introduction. 1933–1941.
Pathologist BSES 1935. Letter to Mr EW Bick, Curator, Botanic Gardens, Brisbane, 31 May. BSES File 507-000. Toads General. 1935–1949.
Pemberton CE (n.d.). Typewritten memoir narrated to 'L.S.M.' In the possession of Michael A Lilly, maternal grandson of Cyril E Pemberton, Honolulu.
Pemberton CE 1921a. Letter to H Agee, 26 January. Sydney. HARC-HSPA records, Honlulu.
Pemberton CE 1928. Letter to AJ Mangelsdorf, 28 June. HARC-HSPA records, Honlulu.
Pemberton CE 1929. Diary 13 June. In the possession of Mary Larson, nee Pemberton, daughter of Cyril E Pemberton, Seattle.
Pemberton CE 1932a. Annual Report of Committee in Charge of the Experiment Station. Hawaiian Sugar Planters' Association. Year ending 30 September.
Pemberton CE 1932b. Chairman's remarks. Section E: Insect pests of sugar cane. Biological control. Proceedings of the fourth congress of International Sugar Cane Technologists, San Juan, Puerto Rico.
Pemberton CE 1932c. Letter to OH Swezey, 23 March. Bishop Museum, Honolulu, MS Group 91.

Works cited

Pemberton CE 1932d. Radiogram to LCD Experiment Honolulu, RCA Communications Inc., 5 April. Bishop Museum, Honolulu, MS Group 91.

Pemberton CE 1932e. The present status of sugar cane insects in Hawaii. Proceedings of the Association of Hawaiian Sugar Cane Technologists, Honolulu, October.

Pemberton CE 1932f. Diary 3 March. In the possession of Mary Larson, nee Pemberton, daughter of Cyril E Pemberton, Seattle.

Pemberton CE 1932g. Diary 22 March. In the possession of Mary Larson, nee Pemberton, daughter of Cyril E Pemberton, Seattle.

Pemberton CE 1932h. Diary 22 April. In the possession of Mary Larson, nee Pemberton, daughter of Cyril E Pemberton, Seattle.

Pemberton CE 1935a. Letter to RW Mungomery, 14 October. BSES File 507-0000. Toads General. 1935–1949.

Pemberton CE 1935b. Recent control measures against *Anomala orientalis* in Hawai'i. Proceedings of the fifth congress of the International Society of Sugar Cane Technologists, Brisbane, pp. 591–94.

Pemberton CE 1935c Diary 8 May. In the possession of Mary Larson, nee Pemberton, daughter of Cyril E Pemberton, Seattle.

Pemberton CE 1936. Letter to AF Bell, 11 May. BSES File 507-0000. Distribution. 1935–1950.

Pemberton CE 1937a. Letter to HL Lyon, April 5 1937. Bishop Museum, Honolulu, MS Group 91.

Pemberton CE 1937b. Diary 25 September. In the possession of Mary Larson, nee Pemberton, daughter of Cyril E Pemberton, Seattle.

Perkins RCL 1943. The early works of Albert Koebele in Hawai'i. In PH Timberlake, *The Coccinellidae or ladybeetles of the Koebele Collection*, part 1. Bulletin of the Experiment Station of the Hawaiian Sugar Planters' Association. Online: www.nhm.ac.uk/resources/research-curation/projects/chalcidoids/pdf_x/perkin925.pdf. Retrieved 14 August 2013.

Perkins RCL & Kirkaldy GW 1907. Leaf-hoppers and parasites of leaf-hoppers. Experiment Station of the Hawaiian Sugar Planters' Association Division of Entomology, Honolulu, Bulletin 4.

Porto Rican Express Company 1932. 29 March, 6 April, 12 April. Leanse Las Condiciones de Este Recibo. Bishop Museum, Honolulu, MS Group 91.

Premier of Queensland 1935. Letter to the prime minister of the Commonwealth, 2 December. Queensland State Archives item ID 863194 (PRE/A1134.) General correspondence 07490 of 23 December.

Prime minister of Australia 1935. Letter to the Premier of Queensland, 17 December. Queensland State Archives item ID 863194 (PRE/A1134). General correspondence 07490 of 23 December.

Release notes 1935–1950. BSES File 507-000. Distribution. 1935–1950.

Riley CV 1892. *Directions for collecting and preserving insects*. Bulletin 39, Smithsonian Institution, United States National Museum.

Rivett D 1935. Letter to W Kerr, 18 November. BSES File 507-0000. Distribution. 1935–1950.

Secretary Australian Country Party, Queensland 1972. Letter to the Hon. WAR Rae, Minister for Local Government and Electricity, 6 April. BSES File 507-0000. Toads Distribution. 1961–1981.

Secretary Queensland Department of Agriculture and Stock 1935. Letter to the Premier of Queensland, 22 November. Queensland State Archives item ID 863194 (PRE/A1134). General correspondence 07490 of 23 December.

Shine R 2012. Personal communication, December.

Shire Clerk, Isis Shire Council 1928. Letter to Director, Sugar Experiment Station, 27 February. RW Mungomery Personnel File, BSES.

Swezey OH 1932. Letter to CE Pemberton. 19th April. Bishop Museum, Honolulu, MS Group 91.

Town Clerk and city manager, Maryborough City Council 1949. Letter to the secretary, Department of Agriculture and Stock, 21 December. BSES File 507-000. Toads General. 1935–1949.

Tucker RWE & Wolcott GN 1935. Parasite introductions: Barbados and Puerto Rico. Proceedings of fifth congress of International Sugar Cane Technologists, Brisbane, pp. 398–105.

Undersecretary Queensland Department of Agriculture and Stock 1935. Letter to Director General of Health, 20 November. BSES File 507-0000. Distribution. 1935–1950.

Undersecretary Queensland Department of Agriculture and Stock 1936. Letter to Director General of Health, 8 July. BSES File 507-0000. Distribution. 1935–1950.

Wassersug RJ 2008. Adjunct Professor in the Department of Urologic Sciences, University of British Columbia. Personal communication.

Whish CB 1862a. Diaries, 11 October. Queensland State Library, John Oxley Library. Box 8600. OM 65-33/8.

Whish CB 1862b. Diaries, 25 October 1862. Queensland State Library, John Oxley Library. Box 8600. OM 65-33/8.

Whish CB 1863a. Diaries, 29 January. Queensland State Library, John Oxley Library. Box 8600. OM 65-33/9.

Whish CB 1863b. Diaries, 3 April. Queensland State Library, John Oxley Library. Box 8600. OM 65-33/9.

White EA 1940. Letter to RW Mungomery, 2 February. BSES File 507-000. Toads General. 1935–1949.

White EA 1947. Letter to RW Mungomery, 21 July. BSES File 507-000. Toads General. 1935–1949.

Wolcott GN 1935. The white grub problem in Puerto Rico. Proceedings of the fifth congress of the International Sugar Cane Technologists, Brisbane, pp. 445–56.
Wolcott G N 1951 The present status of economic entomology in Puerto Rico. Bulletin 99 University of Puerto Rico Agricultural Experiment Station, Rio Piedras, Puerto Rico.

Secondary sources

A History of the Hope Family, Hopetoun (n.d.). Hopetoun House Preservation Trust. Online: www.hopetoun.co.uk/History-of-the-Hope-Family.html. Retrieved 6 August 2013.
Allen A 1995. *A dictionary of Sussex folk medicine*. Countryside Books, Berkshire.
Allsopp PG 2001. The research process – canegrub management in the Australian industry. *Proceedings Australian Society of Sugar Cane Technologists*, 23, pp. 9–16.
Anon 1882. Louisiana sugar and rice crops. Planters' Labour and Supply Company of the Hawaiian Islands. *The Planters' Monthly*, 1, p. 220.
Anon 1900a. *Hawaiian Planters' Monthly*, 19(1) (15 January). Hawaiian Sugar Planters' Association, Honolulu.
Anon 1900b. *The Louisiana Planter and Sugar Manufacturer*, 24(14), 7 April.
Anon 1918. *Hawaiian Planters' Record*, 19(1), July 1918.
Anon 1930. *The story of Queensland sugar*. Queensland Cane Growers' Council, p. 20.
Anon 1933. *Sugar annual*. North Queensland Register, pp. 5–6.
Anon 1936. *The Australian cane sugar industry*. The Sugar Industry Organisations. Queensland Producer Pty Ltd, Brisbane, p. 48.
Anon 1979. Queensland cane toad hops into the export trade. *Overseas Trading: Journal of Australian Department of Trade and Resources*, October, p. 11.
Anon 2000. Harvard 'Honours' Dalhousie researcher with international award. *Connection*, October/November 2000. Online: communications.medicine.dal.ca/connection/octnov2000/harvard.htm. Retrieved 6 August 2013.
Augeron M & Vidal L 2007. Creating colonial Brazil: the first donatary captaincies, or the system of private exclusivity (1534–1549). In Roper LH & van Ruymbeke B (eds) *Constructing modern empires: proprietary ventures in the Atlantic World, 1500–1750*. Koninklijke Brill NV, Leiden.
Australian Government, Department of Agriculture, Fisheries and Forestry (n.d.). Import Conditions. Online: www.aqis.gov.au/icon32/asp/ex_querycontent.asp. Retrieved 12 August 2013.

Australian Government Department of Health and Ageing. Office of the Gene Technology Regulator (n.d.). Online: www.ogtr.gov.au. Retrieved 6 September 2013.

Autour de la 'Mouffe', le turbulent Fabourg St Marcel, Paris Révolutionnaire (n.d.). Online: www.parisrevolutionnaire.com/spip.php?article2708. Retrieved 14 August 2013.

Bancroft HH 1888. *The works of Hubert Howe Bancroft. Volume XXIII. History of California Vol VI 1848–1859.* The History Company, San Francisco. Online: archive.org/details/cihm_14174 Retrieved 8 September 2013.

Barber L 1980. *The heyday of natural history 1820–1870.* Jonathan Cape, London.

Beckman C & Shine R 2009. Impact of invasive cane toads on Australian birds. *Conservation Biology* 23, pp. 1,544–49.

Begg G, Walden D & Rovis-Hermann J 2002. Report on the joint *eriss*/PAN cane toad risk assessment field trip to the Katherine/Mataranka and Borroloola regions. Office of the Supervising Scientist. Internal report 389. July 2002.

Bell AF 1956. *The story of the sugar industry in Queensland.* John Thomson Lecture, 1955. University of Queensland Press, St Lucia.

Bethell, JT (n.d.). 'A splendid little war': Harvard and the commencement of a new world order, *Harvard Magazine.* Online: harvardmagazine.com/1998/11/war.html. Retrieved 15 August 2013.

Bettinger HF 1950a. Pregnancy tests. *The Medical Journal of Australia*, 37, 15 April, pp. 504–07.

Bettinger H F 1950b Pregnancy Tests. *The Medical Journal of Australia.* 37, 10 June, p. 782.

Bettinger HF & O'Loughlin L 1950. The use of the male toad *Bufo marinus* for pregnancy tests. *The Medical Journal of Australia*, 37, 8 July, pp. 40–42.

Bianchi FA 1977. Cyril Eugene Pemberton, 1886–1975: a biographical sketch. *Proceedings, Hawaiian Entomological Society,* 22, pp. 417–41.

Black WG 1883. *Folk medicine: a chapter in the history of culture.* Folk-lore Society, London.

Boland CRJ 2004. Introduced cane toads are active nest predators and competitors of rainbow bee eaters: observational and experimental evidence. *Biological Conservation,* 120, pp. 53–62.

Brandes EW 1929. Into primeval Papua by seaplane. *National Geographic.* 56, pp. 253–332.

Buhôt J 1864. On the cultivation of the sugar cane, *The Guardian* (Brisbane). State Library of Queensland. John Oxley Library, Brisbane.

Burgess GHO 1967. *The curious world of Frank Buckland.* London: Baker.

Burns GD 2010. *Tanksinker.* Privately published. Online: sites.google.com/site/tanksinker/Home/max-burns-tanksinker. Retrieved 9 August 2013.

Works cited

Cactoblastis Memorial (n.d.). Monument Australia. Online: monumentaustralia.org.au/themes/disaster/plagues/display/91971-cactoblastis-memorial. Retrieved 6 September 2013.

Caltagirone LE & Doutt RL 1989. The history of the vedalia beetle importation into California and its impact on the development of biological control. *Annual Review of Entomology*, 31, pp. 1–16.

Cane toad evolution, Cane toads in Oz (n.d.). Online: www.canetoadsinoz.com/cane_toad_evolution.html. Retrieved 12 August 2013.

Cane toad threat to Lake Eyre Basin 2010. Radio program, ABC Radio. 22 May.

Carson R 1968(1962). *Silent spring*. Penguin Books Ltd, Middlesex.

Catling PC, Hertog A, Burt RJ, Wombey JC & Forrester RI 1999. The short-term effect of cane toads (*Bufo marinus*) on native fauna in the Gulf Country of the Northern Territory. *Wildlife Research*, 26, pp. 161–85.

Charter of the Dutch West India Company, 1621. The Avalon Project. Online: avalon.law.yale.edu/17th_century/westind.asp. Retrieved 15 August 2013.

Commonwealth of Australia 1902 *Sugar Regulations. Regulations to the Excise Act 1901 and the Excise Tariff Act 1902*. Australian Government Publisher.

Coquillett DW 1889. The imported Australian ladybird. *Insect Life*, 2, pp. 70–74.

Costar BJ 2006. William(Bill) Forgan Smith (1887–1953), *Australian Dictionary of Biography*. Online: www.adb.online.anu.edu.au/biogs/A110685b.htm. Retrieved 8 August 2013.

Covacevich J & Archer M 1975. The distribution of the cane toad, *Bufo marinus*, in Australia and its effects on indigenous vertebrates. *Memoirs of the Queensland Museum*, 17, pp. 305–10.

Crook AH 1932. The insect menace – book review. *The Hong Kong Naturalist*, vols 3–4, May.

Cunneen C 2006. Hopetoun, seventh Earl of (1860–1908), *Australian Dictionary of Biography*. Online: adb.anu.edu.au/biography/hopetoun-seventh-earl-of-6730. Retrieved 15 August 2013.

Darwin C 2006(1859). On the origin of species. Facsimile in RO Wilson (ed.) *From so simple a beginning*. Norton and Company, London.

Debach P 1974. *Biological control by natural enemies*. Cambridge University Press, Cambridge.

Dexter RR 1929b. The histogenesis and some points in the gross anatomy of the oviduct of the chick. Master of Science thesis, University of Chicago, Illinois.

Dodd AP 1929. The progress of biological control of prickly-pear in Australia. Commonwealth Prickly-pear Board, Brisbane.

Donlan J, Greene H W, Berger J, Bock CE, Bock JH, Burney DA, Estes JA, Foreman D, Martin PS, Roemer GW, Smith FA & Soulè ME. 2005. Re-wilding North America, *Nature*, 436(18).

Doody JS, Green B, Rhind D, Catellano CM, Sims R & Robinson T 2009. Population-level declines in Australian predators caused by an invasive species. *Animal Conservation*, 12, pp. 46–53.

Doody JS, Green B, Sims R, Rhind D, West P & Steer D 2006. Indirect impacts of invasive cane toads (*Bufo marinus*) on nest predation in pig-nosed turtles (*Carettochelys insculpta*). *Wildlife Research*, 33, pp. 349–54.

Dorrance WH 2001. *Sugar islands*. Mutual Publishing, Honolulu.

Dubey S & Shine R 2008. Origin of the parasites of an invading species, the Australian cane toad (*Bufo marinus*): are the lungworms Australian or American? *Molecular Biology* 17, 4, pp. 418–24.

Easteal S 1981. The history of introductions of *Bufo marinus* (Amphibia: Anura); a natural experiment in evolution. *Biological Journal of the Linnean Society*, 16, pp. 93–113.

Easton E 2008. Kershaw, John Crampton Wilkinson. In JL Capinera, *Encyclopedia of Entomology*. Kluwer Academic Press, Dordrecht.

Easton ER 2007. Exploits of some famous entomologists of the Hawaiian Entomological Society. *Proceedings of the Hawaiian Entomological Society*, 39, pp. 153–56.

Ehrlich P, Dobkin DS & Wheye D 1988. Avian invaders. Online: www.stanford.edu/group/stanfordbirds/text/essays/Avian_Invaders.html. Retrieved 12 August 2013.

Estes R & Wassersug R 1963. A Miocene toad from Colombia, South America. *Breviora*, 193, pp. 1–13.

Fitzgerald JF 1944. Early days of the sugar industry in Queensland. *Queensland Geographical Journal*, 49, pp. 68–80.

Fitzgerald R 1984. *From 1915 to the early 1980s: a history of Queensland*. University of Queensland Press, St Lucia.

Fitzgerald R, Megarrity L & Symons D 2009. *Made in Queensland*. University of Queensland Press, St Lucia.

Freeland W 1985. The need to control cane toads. *Search*, 16, pp. 211–15.

Freeland W 1986. Populations of cane toad, *Bufo marinus*, in relation to time since colonisation. *Australian Wildlife Research* 13, pp. 321–29.

Freeland WJ & Martin, KC 1985. The rate of range expansion by *Bufo marinus* in Northern Australia, 1980–84. *Australian Wildlife Research*, 12, pp. 555–59.

Froggatt W 1936b. The introduction of the giant American toad, *Bufo marinus*, into Australia, *Australian Naturalist*, 9(7).

Frogwatch (n.d.) Online: www.frogwatch.org.au. Retrieved 9 August 2013.

Funasaki GY, Po-Yung L, Nakahara LM, Beardsley JW & Ota AS 1988. A review of biological control introductions in Hawaii: 1890 to 1985. *Proceedings Hawaiian Entomological Society*, 28, pp. 105–60.

Works cited

Gill JCH & Lloyd, PL 2006. Veitch, Robert (1890–1972), *Australian Dictionary of Biography*. Online: www.adb.online.anu.edu.au/biogs/A160537b.htm. Retrieved 8 August 2013.

Global Invasive Species Database (n.d.). Online: www.issg.org/database/species/search.asp?st=100ss&fr=1&str=&lang=EN. Retrieved 6 August 2013.

Glubb J 1978. *Into battle: a soldier's diary of the Great War*. Cassell, London.

Goldgar A 2007. *Tulipmania: money, honor, and knowledge in the Dutch golden age*. University of Chicago Press, Chicago.

González-Bernal, E., Greenlees, M. J., Brown, G. P. and Shine, R. (2013), Interacting biocontrol programmes: invasive cane toads reduce rates of breakdown of cowpats by dung beetles. *Austral Ecology*. doi: 10.1111/aec.12028

Grahame K 1995(1908). *The wind in the willows*. New York, Charles Scribner's Sons. Online: etext.virginia.edu/toc/modeng/public/GraWind.html. Retrieved 16 August 2013.

Grammer AR 1947. A history of the Experiment Station of the Hawaiian Sugar Planters' Association 1895–1945. *The Hawaiian Planters' Record*, 51, pp. 177–228.

Graves A 1993. *Cane and labour. The political economy of the Queensland sugar industry, 1862–1906*. Edinburgh University Press, Edinburgh.

Graves R 2009(1929). *Goodbye to all that*. Penguin Books, London. Re-published by Penguin Group (Australia).

Griggs P 2005. Entomology in the service of the state: Queensland scientist and the campaign against cane beetles, 1895–1950. *Historical Records of Australian Science*, 16, pp. 1–29.

Hamilton J 2004. *Goodbye cobber, God bless you*. Pan Macmillan Australia, Sydney.

Harford CF 1911. *Hints on outfit for travellers in tropical countries*. 2nd edition. The Royal Geographical Society, London.

Harper's Weekly, 27 April 1861, vol. v, no. 226, p. 257. Online: www.sonofthesouth.net/leefoundation/civil-war/attack-on-ft-sumter.htm. Retrieved 6 August 2013.

Harvard University Gazette, 28 September 2000. Ig Nobel seeks smartest person in the world. Online: news.harvard.edu/gazette/2000/09.28/ignobel.html. Retrieved 9 September 2013.

Hince B 2011. Paradise, Euroa. Australia's first frog farm. *Locale: The Australasian-Pacific Journal of Regional Food Studies*, 1, pp. 130–55.

Hobbes T 2009(1651). *Leviathan, or the matter, forme, and power of a common-wealth ecclesiastical and civill*. Online: www.gutenberg.org/files/3207/3207-h/3207-h.htm. Retrieved 16 August 2013.

Hopetoun House, Scotland's Places (n.d.). Online: www.scotlandsplaces.gov.uk/
record/rcahms/49127/hopetoun-house/rcahms?item=1167814. Retrieved 14
August 2013.

Horst RG, Hoagland DB & Kilpatrick CW 2001. *The mongoose in the West Indies.*
In CE Woods & FE Sergile (eds). *Biogeography of the West Indies: patterns and
perspectives* (pp. 409-24). CRC Press, Boca Raton.

Howarth FG 1991. Environmental impacts of classical biological control. *Annual
Review of Entomology,* 36, pp. 485-509.

Import Conditions, Department of Agriculture, Fisheries and Forestry (n.d.).
Online: www.aqis.gov.au/icon32/asp/ex_querycontent.asp. Retrieved 12
August 2013.

Joesting E 1972. *Hawaii: an uncommon history.* WW Norton and Co Inc., New York.

Johnson WR 1988. *A documentary history of Queensland.* University of
Queensland Press, St Lucia.

Kearney M, Phillips BL, Tracy K, Christian K, Betts G & Porter WP 2008.
Modelling species distributions without using species distributions: the cane
toad in Australia under current and future climates. *Ecography,* 31, pp.
423-34.

Kerr HW & Bell AF 1939. *The Queensland Cane Growers' Handbook.* Bureau of
Sugar Experiment Stations. Queensland Department of Agriculture,
Government Printer, Brisbane.

Kimberley Toadbusters (n.d.) Online: www.canetoads.com.au. Retrieved 12
August 2013.

Kimberley Toadbusters 2012. Newsletter 46: 22/04/2012. Online:
www.canetoads.com.au/hewslet46.htm. Retrieved 12 August 2013.

Koebele A 1891. Sugar cane insects in New South Wales. *Insect Life,* 4, pp. 385-89.

Koning DAW 1961. *Facsimile of the first Amsterdam pharmacopoeia, 1636.* B de
Graaf, Nieuwkoop.

Kuble A 1879. Le marché aux crapauds. *Le Journal illustré, 7 Septembre.* p. 283.
Translated 2011 by Ms Delphine Swat, Darwin.

Kuykendall RR 1938. *The Hawaiian kingdom,* vol. I, *1778-1854 Foundation and
transformation.* University of Hawaii Press, Honolulu.

Kuykendall RR 1953. *The Hawaiian kingdom,* vol. II, *1854-1874 Twenty critical
years.* University of Hawaii Press, Honolulu.

Ladinsky L 2010. Interview in M Lewis (director) 2010, *Cane toads: the conquest.*
Radio Pictures, Mullumbimby (New South Wales).

Lawrence PO 2000. The pioneer work of George N Wolcott: implications for
US-Caribbean entomology in the 21st century. *Florida Entomologist,* 83, pp.
333-99.

Leeser O 1959. Bufo. *The British Homeopathic Journal,* 48, pp. 176-88.

Works cited

Legislative Assembly of the Northern Territory 2003. Sessional Committee on Environment and Sustainable Development. Issues associated with the progressive entry into the Northern Territory of cane toads, vol. 3. Hansard Transcripts. Public Hearings. October.

Lelong BM 1900. *Culture of the citrus in California*. California State Board of Horticulture, Sacramento.

Letnic M & Ward S 2005. Observations of freshwater crocodiles (*Crocodylus johnstoni*) preying upon cane toads (*Bufo marinus*) in the Northern Territory. *Herpetofauna*, 35, pp. 98–99.

Letnic M, Webb JK & Shine R 2008. Invasive cane toads (*Bufo marinus*) cause mass mortality of freshwater crocodiles (*Crocodylus johnstoni*). *Biological Conservation*, 141, pp. 1,773–82.

Lever C 1992. *They dined on Eland. The story of acclimatisation societies*. Quiller Press, London.

Lever C 2001. *The cane toad. The history and ecology of a successful colonist*. Westbury Academic and Scientific Publishing, Otley, West Yorkshire.

Leverington KC 2006. Mungomery, Reginald William (Reg) (1901–1972), *Australian Dictionary of Biography*. Online: www.adb.online.anu.edu.au/biogs/A150510b.htm. Retrieved 8 August 2013.

Lewis M (director) 1988. *Cane toads: an unnatural history*. Radio Pictures, Mullumbimby (New South Wales).

Lewis M (director) 2010. *Cane toads: the conquest*. Radio Pictures, Mullumbimby (New South Wales).

Lewis S 1989. *Cane toads: an unnatural history*. Dolphin/Doubleday, Sydney.

Licht LE 1967. Death following possible ingestion of toad eggs. *Toxicon*, 5, pp. 141–42.

Ligon R 1657. A true and exact history of the islands of Barbadoes. Online: nationalhumanitiescenter.org/pds/amerbegin/permanence/text1/LigonBarbados.pdf. Retrieved 15 August 2013.

Linnaeus C 1758. *Systema Naturae. Tomus 1*. Editio Decima, Reformata. Holmiae, Laurentii Salvii.

Llewelyn J, Phillips BL, Alford RA, Schwarzkopf L & Shine R 2010. Locomotor performance in an invasive species: cane toads from the invasion front have greater endurance, but not speed, compared to conspecifics from a long-colonised area. *Oecologia*, 162, pp. 343–48.

Lowndes AG (ed.) 1956. *South Pacific enterprise: the Colonial Sugar Refining Company Limited*. Angus & Robertson, Sydney.

Malashichev YB & Wassersug RJ (2004). Left and right in the amphibian world: which way to develop and where to turn? *BioEssays*, 26, pp. 512–22.

Marino Leather Exports (n.d.). Online: www.toadfactory.com. Retrieved 9 August 2013.

McDonald DI 2006. Froggatt, Walter Wilson (1858–1937), *Australian Dictionary of Biography*. Online: www.adb.online.anu.edu.au/biogs/A080608b.htm. Retrieved 15 August 2013.

McKinnon R 2007. Hill, Walter (1819–1904), *Australian Dictionary of Biography*. Online: adb.anu.edu.au/biography/hill-walter-12981. Retrieved 15 August 2013.

Medawar PB 1969. *The art of the soluble*. Pelican Books, Harmondsworth.

Merian MS 1705. *Les insectes de Surinam: metamorphosis insectorum Surinamensium*. Taschen, Hong Kong, 2009.

Minister's decision to develop a threat abatement plan for cane toads. Department of Sustainability, Environment, Water, Population and Communities (n.d.). Online: www.environment.gov.au/epbc/notices/cane-toads-decision.html. Retrieved 12 August 2013.

Molen A 1971. Population and social patterns in Barbados in the early eighteenth century. *The William and Mary Quarterly, Third Series*, 28, pp. 287–300.

Morton C 1995, *By strong arms*. The Mulgrave Central Mill Company Limited, Gordonvale (Queensland).

Mungomery RW 1934b. The control of insect pests of sugar cane. *The Cane Growers' Quarterly Bulletin*, 2, pp. 1–8.

Mungomery RW 1935e. The Giant American Toad (*Bufo marinus*). *Cane Growers' Quarterly Bulletin*, 1 July.

Mungomery RW 1950. A review of sugar cane entomological investigations. In *Fifty years of scientific progress*. Bureau of Sugar Experiment Stations. Government Printer, Brisbane.

Müsch I, Willmann R & Rust J 2001. *Albertus Seba's cabinet of natural curiosities: Locupletissimi rerum naturalium thesauri 1734–1765*. Taschen, Köln.

Nairn ME, Allen PG, Inglis AR & Tanner C 1996. *Australian quarantine; a shared responsibility*. Department of Primary Industry, Canberra.

Naitoh T & Wassersug R 1996. Why are toads right handed? *Nature*, 380(6569), p. 30.

New Scientist. 24 March 1990.

Northern Territory Department of Land Resource Management (n.d.). Feral buffalo. Online: www.lrm.nt.gov.au/feral/buffalo#.UivRXRYn3iw. Retrieved 8 September 2013.

Ormiston House Friends and Advisers Committee. Ormiston House (n.d.). Online: ormistonhouse.org.au. Retrieved 8 September 2013.

Owen W 1966. Anthem for doomed youth. *The Poems of Wilfred Owen*. Chatto and Windus, London.

Parks Australia: Kakadu National Park (n.d.). Australian Government Department of Sustainability, Environment, Water, Population and Communities. Online: www.environment.gov.au/parks/kakadu/. Retrieved 8 September 2013.

Works cited

Peacock D & Abbott I 2010. The mongoose in Australia: failed introduction of a biological control agent. *Australian Journal of Zoology,* 58, pp. 205–27.

Pemberton CE 1921b. The fern weevil parasite. *The Hawaiian Planters' Record,* 25, pp. 196–201.

Pemberton CE 1921c. The fig wasp in its relation to the development of fertile seed in the Moreton Bay fig. *The Hawaiian Planters' Record,* 24, pp. 297–319.

Pemberton CE 1933. Introduction to Hawaii of the tropical American toad *Bufo marinus. The Hawaiian Planters' Record,* 37, pp. 15–16.

Pemberton CE 1934. Local investigations on the introduced tropical American toad *Bufo marinus. The Hawaiian Planters' Record,* 38, pp. 186–92.

Pemberton CE 1948. History of the Entomology Department Experiment Station 1904–1945, *Hawaiian Planters' Record,* 52(1).

Pemberton CE 1964. Highlights in the history of entomology in Hawaii 1778–1963. *Pacific Insects,* 6, pp. 689–729.

Perkins R C L (n.d.). *The Early Work of Albert Koebele in Hawai'i* Untethered Extract, Natural History Museum, London. avalanche.nhm.ac.uk/resources/research-curation/projects/chalcidoids/pdf/Perkin925.pdf. Viewed 16th May 2011.

Perkins RCL 1913. *Fauna Hawaiiensis or the Zoology of the Sandwich (Hawaiian) Isles. Introductory essay on the* fauna, vol. 1, part 6. Cambridge University Press, Cambridge, pp. xv–ccxxviii. Online: hbs.bishopmuseum.org/pubs-online/fh.html. Retrieved 8 September 2013.

Phillips BL, Brown GP & Shine R 2003. Assessing the potential impact of cane toads on Australian snakes. *Conservation Biology,* 17, pp. 1,738–47.

Pramuk J, Robertson T, Sites Jr JW & Noonan BP 2008. Around the world in 10 million years: biogeography of the nearly cosmopolitan true toads (Anura: Bufonidae). *Global Ecology and Biogeography,* 17, pp. 72–83.

Pugh TP 1863. *Pugh's Queensland almanac, directory and law calendar 1863.* Theophilus P Pugh, Brisbane.

Queensland Government (n.d.). Department of Environment and Resource Management. Riversleigh Section, Boodjamilla (Lawn Hill) National Park. Online: www.nprsr.qld.gov.au/parks/boodjamulla-lawn-hill/. Retrieved 6 September 2013.

Queensland Government Gazette, 28 March 1938.

Queensland Parliament House. Member Information (n.d.). Online: parliament.qld.gov.au/view/historical/documents/memberBio/MooreA.htm. Retrieved 7 August 2013.

Readers Digest, October 1980.

Remarque EM 1993(1929). *All quiet on the Western Front (Im Westen nichts Neues).* Picador, Pan Books Ltd, London.

Riley CV 1869. Toads in gardens. *American Entomologist,* 2(50).

Roe M 2006. Cumpston, John Howard Lidgett (1880-1954), *Australian Dictionary of Biography*. Online: www.adb.online.anu.edu.au/biogs/A080194b.htm. Retrieved 8 August 2013.

Rolls EC 1969. *They all ran wild*. Angus & Robertson, Sydney.

Schama S 1997. *The embarrassment of riches. An interpretation of Dutch culture in the Golden Age*. Vintage Books, New York.

Schomburgk RH 1848. *The history of Barbados*. Longman, Brown Green and Longmans, London.

Schwartz A & Henderson RW 1991. *Amphibians and reptiles of the West Indies: descriptions, distributions, and natural history*. University of Florida Press, Gainesville.

Scott E 1941. *Official history of Australia in the war of 1914-1918*, vol. 11, *Australia during the war*. Online: www.awm.gov.au/histories/first_world_war/volume.asp?levelID=67897/. Retrieved 7 August 2013.

Shanmuganathan T, Pallister J, Doody S, McCallum H, Robinson T, Sheppard A, Hardy C, Halliday D, Venables D, Voysey R, Strive T, Hinds L and Hyatt A 2010. Biological control of the cane toad in Australia: a review. *Animal Conservation*, 13, pp. 16-23.

Sharp D (ed.) 1913. *Fauna Hawaiiensis or the zoology of the Sandwich (Hawaiian) Isles*. Preface, vol. 1, part 6. Cambridge University Press. Online: hbs.bishopmuseum.org/pubs-online/fh.html. Retrieved 8 September 2013.

Shine R (n.d.). Feasibility of cane toad control. Cane Toads in Oz. Online: www.canetoadsinoz.com/controlfeasibility.html Viewed. Accessed 9 August 2013.

Shine R 2007. Mr Toad comes to Darwin: an evolutionary perspective on the cane toad invasion. Australian Academy of Sciences Public Lecture Series 2006-07. Online: science.org.au/events/publiclectures/os/shine.html. Retrieved 9 August 2013.

Shine R 2010. The ecological impact of invasive cane toads (*Bufo marinus*) in Australia. *The Quarterly Review of Biology*, 85, pp. 253-91.

Slade N & Wassersug RJ 1975. On the evolution of complex life cycles. *Evolution*, 29, pp. 568-71.

Sloane H 1725. *A voyage to the islands Madera, Barbadoes, Nieves, St Christophers, and Jamaica*, vol II. Online: www.biodiversitylibrary.org/bibliography/642#/summary. Retrieved 31 July 1913.

Smith JG & Phillips BL 2006. Toxic tucker: the potential impact of cane toads on Australian reptiles. *Pacific Conservation Biology*, 12, pp. 40-49.

Stokes JL 2004(1846). *Discoveries in Australia; with an account of the coasts and rivers explored and surveyed during the voyage of H.M.S. Beagle, in the years 1837-38-39-40-41-42-43*. Online: www.gutenberg.org/files/12115/12115.txt. Retrieved 9 August 2013.

Works cited

Stop the toad (n.d.). www.stopthetoad.org.au. Retrieved 13 October 2013.
Stress helps cane toads' desert invasion 2013. *News in Science*. ABC Science. 14 August.
Swan C 2005. *Art, science and witchcraft in early modern Holland*. Cambridge University Press, New York.
Sweetman HL 1958. *The principles of biological control*. Wm. C Brown Company, Dubuque (Iowa).
Sydney Morning Herald, 4 May 1928.
Sydney Morning Herald, 20 September 1928.
Sydney Morning Herald, 29 November 1935.
Sydney Morning Herald, 22 September 1937.
Sydney Morning Herald, 2 August 1947.
Sydney Morning Herald, 19 April 1996.
Sydney Morning Herald, 7 April 2001.
Sydney Morning Herald, 17 September 2007.
The Adelaide Observer, 13 October 1888.
The Argus (Melbourne), 25 June 1935.
The Brisbane Courier, 3 May 1932.
The butterfly man of Kuranda (n.d.). Queensland Museum. Online: www.southbank.qm.qld.gov.au/Events+and+Exhibitions/Exhibitions/2010/02/Butterfly+Man+of+Kuranda. Retrieved 7 August 2013.
The Canberra Times, 29 May 1947. Beekeepers oppose toad introduction.
The Courier Mail, 26 September 1933.
The Courier Mail, 29 November 1935.
The Courier Mail, 30 November 1935.
The Courier Mail, 2 December 1935.
The Courier Mail, 14 May 1979.
The Courier Mail, 12 November 1981.
The Courier Mail, 18 January 1986.
The Daily Telegraph (Sydney), 1 June 1937.
The Giant Toad 1949. *The Country Hour*, radio program, ABC Radio. 18 March.
The Hawaiian Commercial and Sugar Company on Maui (n.d.). Online: www.hcsugar.com/. Retrieved 9 September 2013.
The Independent. 23 August 1994. Obituary: Rear-Admiral Sir Hugh Janion. Online: www.independent.co.uk/news/people/obituary-rearadmiral-sir-hugh-janion-1378144.html. Retrieved 9 August 2013.
The Land, 30 August 1947.
The New York Times, 28 December 1921.
The New York Times, 22 February 1932.

The New York Times, 27 July 1981. Royal wedding gifts: extraordinary and ordinary.
The Queenslander, 28 November 1935.
The Queenslander, 19 August 1937.
The Queenslander, 26 January 1938.
The Rewilding Institute (n.d.). Online: www.rewilding.org/pleistocene_rewilding.html. Retrieved 12 August 2013.
Threatened Species Scientific Committee (TSSC) 2005. The biological effects, including lethal toxic ingestion, caused by Cane Toads (*Bufo marinus*) – 12 April 2005. Department of the Environment, Water, Heritage and the Arts. Online: www.environment.gov.au/biodiversity/threatened/ktp/cane-toads.html. Retrieved 12 August 2013.
Time, 5 August 1974.
Turvey ND 2010. The toad's tale: a true fable of science and society. In K Weber (ed.), *Cane toads and other rogue species*. Public Affairs, New York. pp. 3–18.
Tyler MJ 1976. *Frogs*. William Collins, Sydney.
Tyler MJ, Wassersug RJ & Smith B 2007. How frogs and humans interact: influences beyond habitat destruction, epidemics and global warming. *Applied Herpetology*, 4, pp. 1–18.
United States Department of Agriculture (n.d.). Charles Valentine Riley Special Collection. National Agricultural Library, Special Collections. Online: specialcollections.nal.usda.gov/guide-collections/charles-valentine-riley-collection. Retrieved 6 September 2013.
van Beurden EK & Grigg G 1980. An isolated and expanding population of the introduced toad *Bufo marinus* in New South Wales. *Australian Wilderness. Res.*, 7, pp. 305–10.
van Dam RA, Walden DJ & Begg GW 2002. *A preliminary risk assessment of cane toads in Kakadu National Park*. Scientist Report 164, Supervising Scientist, Darwin NT.
Waite FC 1901. *Bufo agua* in the Bermudas. *Science*, 13(322), pp. 342–43.
Wallace AR 2008(1890). *The Malay Archipelago*. 10th edition. Facsimile edition. Periplus Editions, Singapore.
Walsh DB & Riley CV 1869. Do toads eat worker bees? *American Entomologist*, 1, p. 144.
Wannan B 1970. *Folk medicine*. Hill of Content Publishing Co Pty Ltd, Melbourne.
Ward-Fear G, Brown GP, Greenlees M J, Shine R 2009. Maladaptive traits in invasive species: in Australia, cane toads are more vulnerable to predatory ants than are native frogs. *Functional Ecology*, 23, pp. 559–68.
Ward-Fear G, Brown PG & Shine R 2010. Using a native predator (the meat ant, *Iridomyrmex reburrus*) to reduce the abundance of an invasive species (the

cane toad, *Bufo marinus*) in tropical Australia. *Journal of Applied Ecology,* 47, pp. 273-80.
Warner CD 2006(1870). *Summer in a garden, and Calvin, a study of character.* Online: www.gutenberg.org/files/3135/3135-h/3135-h.htm. Retrieved 14 August 2013.
Wassersug RJ 1971. On the comparative palatability of some dry-season tadpoles from Costa Rica. *American Midland Naturalist,* 86, pp. 101-09.
Wassersug RJ 1975. The adaptive significance of the tadpole stage with comments on the maintenance of complex life cycles in aurans. *American Zoology,* 15, pp. 405-17.
Wassersug RJ 1997. Where the tadpole meets the world – observations and speculations on biomechanical and biochemical factors that influence metamorphosis in anurans. *American Zoology,* 37, pp. 124-36.
Wassersug RJ 2001. Vertebrate biology in microgravity. *American Scientist,* 89, pp. 46-53.
Wassersug RJ, Izumi-Kurotani A, Yamashita M & Naitoh T 1993. Motion sickness in amphibians. *Behavioural and Neural Biology,* 60, pp. 42-51.
Weidman RM 1997. Memorial to Robert W Fields 1920-1995. *Geological Society of America Memorials,* 28, pp. 1-3.
Wharton WJL (ed.) 2012. *Captain Cook's Journal during his first voyage round the world made in H.M. Bark 'Endeavour'?1768-71.* eBooks@Adelaide, Adelaide. Online: ebooks.adelaide.edu.au/c/cook/james/c77j/. Retrieved 12 August 2013.
Williams E 1970. *From Columbus to Castro: the history of the Caribbean 1492-1969.* Andre Deutsch, London.
Willits E 1889. A letter on *Icerya purchasi*. *Insect Life,* 2, pp. 15-17.
Wolcott GN 1948. The insects of Puerto Rico. *The Journal of Agriculture of the University of Puerto Rico,* 32(2), pp. 225-385.
Wolcott GN 1950. The rise and fall of the white grub problem in Puerto Rico. *The American Naturalist,* 84, pp. 183-93.
Wood CT 2007. Buhôt, John (1831-1881). *Australian Dictionary of Biography.* Online: adb.anu.edu.au/biography/buhot-john-3107. Retrieved 14 August 2013.
Young JE, Christian KA, Donnellan S, Tracy CR & Parry D 2005. Comparative analysis of cutaneous evaporative water loss in frogs demonstrates correlation with ecological habits. *Physiological and Biochemical Zoology,* 78, pp. 847-56.
Zimmerman H, Bloem S & Klein K 2004. *Biology, history, threat, surveillance and control of the cactus moth, Cactoblastis cactorum.* IAEA/FAO-BSC/CM. IAE, Austria. Online: www-pub.iaea.org/mtcd/publications/pdf/faobsc_web.pdf. Retrieved 12 August 2013.

Zimmermann HG, Moran VC & Hoffmann JH 2000. The renowned cactus moth, *Cactoblastis cactorum*: its natural history and threat to native *Opuntia* floras in Mexico and the United States of America. *Diversity and Distributions*, 6, pp. 259–69.

Zug GR & Zug PB 1979. The marine toad, *Bufo marinus*: a natural history resumé of native populations. *Smithsonian Contributions to Zoology*, 284, p. 58.

Index

acclimatisation societies, 46, 215. *See also* American Acclimatisation Society; Society for the Acclimatisation of Animals, Birds, Fishes, Insects and Vegetables; Queensland Acclimatisation Society
Agee, Hamilton, 91
American Acclimatisation Society, 212
Archer, Mike, 178–179
armyworms, 6, 119
Aschheim-Zondek test, 167
Australian National Research Council, 144
Australian Sugar Producers' Association, 148, 152
Australian Wattle League, 144

barramundi, 191
bees, 162, 181, 186
Bell, Arthur
 defence of the introduction of the cane toad, 153, 158, 160
 proponent for the introduction of the cane toad to Australia, 92–93, 95–96, 121, 133, 138, 149, 213
 service in the First World War, 74
 support for the White Australia policy, 93, 95
benzene hexachloride (BHC), 161
biological control, xiii, 25, 53, 66, 71, 72, 81, 87, 116, 130, 140, 145, 181, 200, 205, 216
biology of the cane toad, 5, 6, 155, 181
birds, 6, 47, 48, 102, 103, 137, 147, 153, 173, 174, 176, 178, 179, 182, 185, 186, 198, 211–213
blowflies, 146
bluetongue skinks, 198
Bowen, George, 39, 47
Brandes, Elmer, 93–95
breeding the cane toad, 117, 138
British Association for the Advancement of Science, 63
Buckland, Francis Trevelyan, 46
buffalo, feral, 186
Buhôt, John, 42, 47
Bulcock, Frank, 140, 152
Bureau of Sugar Experiment Stations (BSES), 72, 73, 77, 80, 87, 92, 126–134, 140, 143–144, 147–148,

150–152, 155, 157, 163, 204, 208, 213, 214, 216
  recognition of the failure of the cane toad to control the cane grub, 160, 177
Burns, Max, 175
Buzacott, James, 134, 161
Bynoe's gecko, 189
Byrne, Tip, 125

Cactoblastis moth, 84
  introduction to Australia (Queensland), 85, 206
cane beetles and grubs, xii, 48, 73, 77, 125–130, 156, 205
  anomala, 98, 99, 123, 160
  Childers, 80
  French's, 80
  greyback, 78, 160
  May, 98, 102, 106, 107, 109, 110, 113, 114, 160, 208
  negatoria, 80
  rhopaea, 80
  scarab beetle, 48, 79, 98, 102, 110, 121
  southern one-year cane grub, 80
cane flies, 27
cane grubs and beetles in Australia (Queensland), 81, 125–130, 160, 161
the cane toad in Australia
  Adelaide River, 189, 192
  Argyle Dam, 195
  Atherton Tableland, 175
  Boodjamulla, 178
  Borroloola, 180
  Brisbane, 162
  Byron Bay, 164
  Daly River, 187, 188, 192, 210
  Darwin, 193, 197, 200
  Fogg Dam, 189, 191
  Gordonvale, 157, 159, 213
  Humpty Doo, 192
  Innisfail, 161
  Julia Creek, 175–177
  Kakadu National Park, 185–186
  Katherine, 187
  Lake Eyre, 201
  Longreach Lagoon, 194
  Miriam Vale, 163
  New South Wales, 164
  Nitmiluk National Park, 186
  Percy Island, 163
  Sydney, 201
  the Northern Territory, ix, xiv, 180, 182
  the Great Dividing Range, 175
  the Gulf of Carpentaria, 175, 179
  the Kimberley region, 195, 202
  the Top End, 182, 185–202
  Western Australia, xiii, 194
caterpillars, 27, 132
centipedes, 123
Chataway, James, 72
Chinese grasshoppers, 97
Cilento, Raphael, 140
classification, systematic, 19
climate change, 4, 218
cockchafer grubs, 48. *See also* cane beetles and grubs
Colonial Sugar Refiners (CSR), 48
Commonwealth Department of Health, 132, 146, 149, 209
Commonwealth Prickly-pear Board, 81, 85, 130, 137, 211
Commonwealth Scientific and Industrial Research Organisation (CSIRO), 146, 177, 190, 200
Conservation Commission (Northern Territory), 180
control the cane toad, methods used to, 181, 195
Coquillett, Daniel, 55, 60, 61
cottony cushion scale, 53–55, 58, 60

Index

Council for Scientific and Industrial Research (CSIR), 146, 147–148, 181, 214, 216
Covacevich, Jeanette, 179
Crawford, Frazer, 57
Cumpston, John Howard Lidgett, 146, 151, 156, 213

Darwin, Charles, 197
Denison, Fred, 115–118, 122, 134, 196
Department of Agriculture and Stock (Queensland), 143, 148, 152, 162
de Sousa, Martim Afonso, ix, 25, 120
Dexter, Raquel, 107–114, 129, 132, 156, 208, 212
diet of the cane toad, 6, 117, 156, 162, 163, 205, 209, 211
dingos, 182
Dingwall, Graeme, 180
dissecting the cane toad, 108, 165, 169, 169
Dodd, Alan
  role in the control of the prickly-pear cactus, 75, 207
  service in the First World War, 73, 92
  study of cane grubs and beetles, 79–81, 156
Dodd, Frederick, 76
dogs, death due to cane toads, 123, 158, 159, 162, 176, 178, 187, 192, 199
Doherty, William (Bill), 148, 214
Donlan, Josh, 215
Doody, Sean, 187, 188
dung beetles, 7, 177, 206
dunnarts, 174, 176

Easterby, Harry, 93, 96
El Questro Wilderness Park, 195
Emmott, Angus, 202
Environmental Protection and Biodiversity Conservation Act (1999), 200

Estes, Richard, 2
European toad, 8, 28, 148

fascination with the cane toad, 157, 170. *See also* Kev's Travelling Toad Show
fern weevils, 92, 98
ferrets, 145
Fields, Robert (Bob), 1
fig wasps, 89–90, 98
  introduction to Hawai'i, 91
finches, 188
First World War, 73–77, 105, 147
Forgan Smith, William, 138, 152, 214
fossil 41159, 2, 9, 203
fossil RWF191, 2
Freeland, Bill, 114, 180, 197
freshwater crocodiles, 174, 178, 183, 186, 188, 191, 194, 198
Froggatt, Walter Wilson, 89, 91, 149
  opposition to the introduction of the cane toad, 144–147, 211, 216, 218
  prediction of the spread of the cane toad, 153–156, 202
  relationship with Cyril Pemberton, 89–91, 96, 143, 144, 156
Frogwatch NT, 193

Garrett, Peter, 200
Gene Technology Act (2000), 216
genetically modified organisms (GMOs), 216
Gilbert's dragon, 182, 188
Girault, Alexandre, 77
goannas, 174, 178, 179, 181, 182, 186, 191, 198, 202, 204
  yellow-spotted goanna, 187, 188
Gould League of Bird Lovers of New South Wales, 144
Graham, Arthur, 150, 214
Grahame, Kenneth, 8

Great Depression, 96, 99
Grieg, Elvie, 165

The Hawaiian Sugar Planters' Association (HSPA), 63, 69, 72, 87, 91, 92, 98, 116–116, 119, 121, 130, 138, 143, 155, 196, 204, 208, 209, 213, 214, 216
  Experiment Station, 63, 66, 69, 71, 72, 88, 93
Hill, Walter, 39–42
Hope, Louis, 38–39, 42, 50, 79, 139
horn flies, 206
Howard, Leland, 64

Ig Nobel prize, 5
Illingworth, James, 77, 79–81, 156
integrated pest management (IPM), 207
International Society for Sugar Cane Technologists
  fifth congress, Queensland 1935, 130, 133, 138–141, 143
  fourth congress, Puerto Rico 1932, 96, 99
introduction of the cane toad to Australia: Queensland, xiii, 75, 129, 130–141, 149–150. *See also* the cane toad in Australia
Barbados, 29
Egypt, 121
Fiji, 120
Formosa (now Taiwan), 120
Guam, 122
Jamaica, 28
Japan, 123
Martinique, 29
Mauritius, 122
Micronesia, 206
New Guinea, 95, 120–121
the Philippines, 119

Puerto Rico, 9, 87, 160
the US (Louisiana), 122
the US (Florida), 122, 173
the US (Hawai'i), 9, 87, 99, 115–118, 160

Jeswiet, Jacob, 94

Kakadu National Park, 185. *See also* the cane toad in Australia: Kakadu National Park
Kearney, Mike, 200
Kerr, William (Bill), 129–130, 143, 147–149, 152, 178
  defence of the introduction of the cane toad, 154
Kershaw, John, 69, 70
Kev's Travelling Toad Show, 171
killing cane toads, 157, 161, 169–171, 179, 192–195, 199. *See also* taxidermy; toadbusting
Kimberley Toadbusters (KTB), 194
King, Norman, 177
Koebele, Albert
  ladybird fantasy, 116
  proponent of biological control, 53, 72, 73
  proponent of the introduction of the cane toad, 80, 164
  study of the cane grub and beetle, 48–49, 55–67
Kreuger, John, 170

Ladinsky, Kevin, 170
ladybird fantasy, 61, 116
ladybirds, 53, 58–61, 88, 207
  introduction to Hawai'i, 62
  introduction to Los Angeles, 60
Lantana weed, 66, 163, 206
larrid wasps, 98
Lennox, Colin, 120

Index

Letnic, Mike, 194
Lewis, Mark, 165
lilypad frogs, 192
Linnaeus, Carolus, 19–22, 204
Linnean Society of New South Wales, 144–146
Little Mulgrave River, release of cane toads at, 137, 149–150
lizards, 182
locust swarms, 171
Lucas, Mildred, 120
lungworms, 196–197, 204
Lyons, Joseph, 151–152, 214

Magdalena Valley, 1–3, 112, 203
Mangelsdorf, Al, 94
Maxwell, Walter, 72
McDougall, William, 178
meat ants, 198, 204
 red meat ants, 145
Merian, Anna Maria Sibylla, 19
Merten's water monitor, 188
metamorphosis of tadpoles into toads, 4
mice, 167
mirid bugs, 71, 98
Mitchell's monitor, 188
mole crickets, 92, 98
mongoose, 47, 145, 206
monitor lizards, 206
Moore, Arthur, 96
Muir, Frederick, 67, 69–72, 87, 89
Mungomery, Reg
 defence of the introduction of the cane toad, 153, 157, 160, 162
 initial reluctance to introduce the cane toad to Australia, 133, 156
 proponent of the introduction of the cane toad, 126–141, 144, 149, 196
 release of cane toads in Queensland, 137

muscovado, 26
myths about the cane toad, 5, 17, 164
myxomatosis, 181, 206

naming the cane toad, 164, 204
native environment of the cane toad, 5, 6, 7
Naturalists' Society of New South Wales, 144, 154
negative impact of the introduction of the cane toad (Australia), 161. *See also* toxicity of the cane toad
 anticipation of, 144–156
 on cattle, 176
 on Indigenous communities, 186, 187, 199
 on native fauna, 147, 153, 171, 173–179, 182–183, 186, 212
 on the food chain, 174, 186
 Roper River region study, 183
Northern Territory Parks and Wildlife, 194

Oahu Sugar Company, 99, 116, 117
Oakley, RG, 121
Office of the Gene Technology Regulator (OGTR), 216
'Olympic Village' effect, 197
origins of the cane toad, 1–4
Owen, Richard, 46

palm weevils, 27
Peck, Richard, 94
Pemberton, Cyril
 biography, 87
 defence of the introduction of the cane toad, 155, 162
 introduction of the cane toad to Hawai'i, 87–99, 115–116, 133, 196, 209

245

proponent of the introduction of the cane toad, 147, 154
relationship with Walter Froggatt, 89–91, 96, 143, 144, 154
research into pests of sugar cane, 98
research into the fern weevil and fig wasp, 89, 92
Perkins, Robert, 64, 66, 67, 72, 87, 89
pest control, 28
pets, cane toads as, 164–165, 169
Phillips, Ben, 189, 197
pig-nosed turtle, 187, 188
planigales, 191
Pleistocene Rewilding, 215
predators of the cane toad, 4, 6, 173
pregnancy tests, use of toads in, 166–168
Prickly-pear Board, *see* Commonwealth Prickly-pear Board
prickly-pear cactus, 47, 75, 83–86, 143, 163, 206

quarantine, 35, 95, 98, 132, 146, 148, 213–216
Queensland Acclimatisation Society, 47, 83
Queensland Cane Growers' Council, 148
quolls, 174, 176, 178–179, 182, 186, 199, 202

rabbits, 48, 143, 145, 167, 181, 206
rainbow bee-eaters, 177, 182, 198
rats, 28, 47, 92, 132, 206
raw sugar, *see* muscovado
Riley, Charles Valentine, 49, 53, 62
Rivett, Albert Cherbury David, 146–148, 152, 181, 212, 214, 216
Roth's tree frog, 182
Royal Australian Historical Society, 144

Royal Society of London, 63
Royal Zoological Society of New South Wales, 144

saltwater crocodiles, 180, 191
science, flawed, xiv, 62, 66, 67, 137, 179, 205, 208–211
scoliid wasps, 98, 104, 105, 110
Seba, Albertus, 12, 15–20, 25, 204
Second World War, 73–75, 122
Sharp, David, 63, 64
Shine, Rick, 189, 198, 201. *See also* Team Bufo
slave trade, 14, 18
Smith, Beryl, 186
snakes, 174, 178, 179, 186, 189, 202, 204
  elapid snakes, 198
  keelback snakes, 190, 191
  olive python, 191
  red-bellied black snakes, 189
  slatey-grey snakes, 190
Society for the Acclimatisation of Animals, Birds, Fishes, Insects and Vegetables, 46, 212
stoats, 145
Stop The Toad Foundation, 195
sugar ants, 27
sugar cane
  in Australia, 35, 37, 39, 44, 49
  in Barbados, 28
  in Brazil, ix
  in Cuba, 49
  in Hawai'i, 30, 31–35, 46, 49
  in Jamaica, 28
  in South America, 25, 27, 29
  pests of, xiii, 25, 27. *See also* armyworms; cane beetles and grubs; cane flies; caterpillars; mole crickets; palm weevils; rats; sugar ants; sugar cane weevil borers

Index

sugar cane leaf hoppers, 67, 89, 98
sugar cane weevil borers, 35, 92, 132
   introduction to Hawai'i, 35, 98
sugar, manufacture of, 26, 31, 38, 42, 44, 138, 164
sugar trade, 14, 22, 25, 26
   impact of the American Civil War on, 34–34
   impact of the gold rush on, 32
sweet potato hawk moths, 120
synthetic biology, 217
synthetic insecticides, xii, 61, 126, 161, 207, 216

Tasmanian devil, 179
taxidermy, 170
Team Bufo, 191, 198, 199
Tepper, Otto, 57
Terpstra, Jim, 170
threat posed by the cane toad (in Australia), *see* negative impact of the introduction of the cane toad (Australia)
Threatened Species Scientific Committee (TSSC), 200
Timberlake, Philip, 97
toad markets, 29
toad skins, trade in, 170
toadbusting, 169, 193, 193, 199, 204. *See also* killing cane toads; Kimberley Toadbusters (KTB)
toxicity of the cane toad, xii, 4, 5, 133, 136–138, 158–160, 164, 173–179, 188, 190–192, 198, 211. *See also* negative impact of the introduction of the cane toad (Australia)
domestic animals killed by the cane toad, 159–160, 162, 180. *See also* dogs, death due to cane toads

humans killed by the cane toad, 133, 158, 211
trade in cane toad toxins, 170
Trewenack, John, 58

United Graziers Association, 176
United States Department of Agriculture (USDA), 49, 53, 55, 61, 62, 63, 72, 73, 88, 119, 121, 204, 209, 213

Vedalia beetles, *see* ladybirds
Veitch, Robert, 146, 148
Vesey, Nathaniel, 164
vultures, 146

Waldron, John, 99, 110–111, 114, 116–117, 116, 210, 214
Wassersug, Richard, 2, 4, 8, 189
Waterhouse, Doug, 177
weasels, 145
Wetlands of International Importance, 185
Whish, Claudius Buchanan, 42, 44, 204
White Australia policy, 50, 93, 95, 140
Wildlife Preservation Society of Australia, 144
Williams, FX, 155
Wilson, Brendan, 180
Wilson, Kathy, 186
Windred, GL, 120
wireworms, 122
Wolcott, George, 73, 75, 104–108, 110, 114, 129, 138, 208
wrens, 188

Zug, George, 5

www.ingramcontent.com/pod-product-compliance
Lightning Source LLC
Chambersburg PA
CBHW070640160426
43194CB00009B/1516